Sound Savings

Preserving
Audio
Collections

`1001010010100101010101010` `10101000101010010100`

Proceedings of a symposium sponsored by
School of Information, Preservation and Conservation Studies, University of Texas at Austin
Library of Congress
National Recording Preservation Board
Association of Research Libraries

Austin, Texas

July 24–26, 2003

Edited by Judith Matz

Association of Research Libraries

2004

Sound Savings: Preserving Audio Collections
Judith Matz, Editor

Proceedings of a symposium sponsored by
School of Information, Preservation and Conservation Studies, University of Texas at Austin
Library of Congress
National Recording Preservation Board
Association of Research Libraries

Austin, Texas
July 24–26, 2003

Association of Research Libraries
21 Dupont Circle, NW, Suite 800
Washington, D.C. 20036

ISBN 1-59407-663-4

Design and layout by David S. Noble.

Contents

These papers are available at:
http://www.arl.org/preserv/sound_savings_proceedings/

Preface

William A. Gosling
University Librarian
University of Michigan
Chair, ARL Preservation of Research Library Materials Committee

OVER THE PAST DECADE there has been a growing interest in the preservation needs of sound recordings and other nonprint media. In March 2002, "Redefining Preservation, Shaping New Solutions, Forging New Partnerships," a conference sponsored by the University of Michigan University Library and the Association of Research Libraries (ARL), put audiovisual preservation high on the list of important issues for the preservation community. A year later, many of the key stakeholders in audiovisual preservation met to share their experiences and discuss the challenges ahead, specifically in the area of preserving sound recordings.

"Sound Savings: Preserving Audio Collections," held at the University of Texas at Austin in July 2003, was cosponsored by the School of Information's Preservation and Conservation Studies at the University of Texas at Austin; the Library of Congress; the National Recording Preservation Board; and the Association of Research Libraries. The two-and-a-half day program featured talks by experts on topics ranging from assessing the preservation needs of audio collections to creating, preserving, and making publicly available digitally reformatted audio recordings. As Ellen Cunningham-Kruppa and Mark Roosa noted in their introductory remarks, the symposium brought together an esteemed group of curators, educators, scholars, and practitioners whose papers "represent the fruit of their individual and collective experiences and, as they reveal, institutions are actively involved in all aspects of audio preservation. But there remains serious work ahead."

The final session of "Sound Savings" focused on shaping an applied research and education agenda. Seven steps were defined to move forward; the challenge will be articulating strategies to reach these goals. ARL hopes, by publishing the many outstanding papers presented at the symposium, to advance this agenda. The time has come to expand our present knowledge and work together to meet the challenges of preserving our valuable collections of recorded sound.

Introduction to the Papers

Ellen Cunningham-Kruppa
Assistant Director, Center for the Cultural Record
University of Texas, Austin

Mark Roosa
Director of Preservation
Library of Congress

JUST SEVENTEEN YEARS AGO, the Associated Audio Archives (AAA), a special committee of the Association for Recorded Sound Collections (ASRC), undertook a groundbreaking audio preservation planning study. The National Endowment for the Humanities funded the study, whose goals were to better define and standardize audio preservation methods and practices. The committee, having completed an enormous planning study, pilot project, and finally, a full-fledged project resulting in the Rigler-Deutsch Record Index, a union index to some 615,000 pre-LP commercial recordings held in the then five participating archives, turned its attention to the preservation issues of sound archives. The author of a 1988 briefing of the study, Barbara Sawka, then curator of the Stanford Archive of Recorded Sound, noted that the growing interest in paper preservation in libraries and archives helped to create a more receptive response to the preservation needs of sound recordings and other nonprint media. But, as Ms. Sawka notes in her synopsis of the planning study and we know from experience, sound archivists have had a tough row to hoe to convince colleagues and funding agencies alike to make the pressing needs of audio materials a priority.

In ARSC/AAA's groundbreaking report, *Audio Preservation: A Planning Study* (1988), training and education for sound archivists was discussed at length. Much of the reporting conveyed the technical education needed to operate and maintain audio equipment and to conduct archival sound audio transfer and restoration work.

As the report noted, many of the heads of sound archives in 1988 were subject specialists who had acquired their technical knowledge on the job, not unlike the library preservation field in its nascent years. The report noted that sound archivists or archives administrators had little to choose from in the way of educational programs geared to the requirements of their profession.

Now let's flash forward. In the fifteen years following the publication of the ARSC/AAA study, a number of strides have been made to provide education on the issues of audio preservation. The Society of American Archivists, the American Library Association, the Association for Recorded Sound Archives, the Society of California Archivists, and the North Carolina Preservation Consortium, among other professional groups, have organized workshops and programs for professionals seeking audio preservation training. Educational materials have burgeoned; useful publications on identifying, handling, storing, and reformatting sound media abound in print and on the Web; and many organizations are now involved in sound preservation (http://palimpsest.stanford.edu/bytopic/audio/#organizations). We now also have a number of excellent companies and individuals who provide professional audio reformatting services and consultation, something that has been sorely lacking until recently.

The preservation needs of audio collections have been acknowledged at the national level. Following the National Film Preservation Act of 1996, Congress passed the National Recording Preservation Act in 2000. Subsequently, the National Recording Preservation Board was created and charged with "developing a comprehensive National Recording Preservation Study and Action Plan" to "address issues such as the current state of sound recording archiving, preservation, and restoration activities; research and other activities carried out by or on behalf of the Library of Congress National Audio-Visual Conservation Center at Culpeper, Virginia; the establishment of clear standards for copying old sound recordings; and current laws and restrictions regarding the preservation and use of sound recordings, including recommendations for changes to allow digital access and preservation" (http://www.loc.gov/rr/record/nrpb/nrpb-about.html).

In March 2002, "Redefining Preservation, Shaping New Solutions, Forging New Partnerships," a conference sponsored by the University of Michigan University Library and the Association of Research Libraries (ARL), called attention to the issues of audiovisual preservation by placing them high on the list of important issues for the preservation community. In response to the priorities set forth during the Michigan/ARL conference, the Council on Library and Information Resources (CLIR) is presently conducting a survey of the state of recorded sound in academic libraries. The goals of the survey are to inform decision makers in academic libraries about the state of audio collections, the importance of audio collection for research and teaching, and how to lower the barriers to access (http://www.clir.org/pubs/issues/issues32.html#audio).

In short, the profession has raised awareness about the cultural and sociological importance of sound, and has made sound accessible via intellectual control and preservation efforts.

During the June 2002 Annual Meeting of the American Library Association, we began to think about and plan a gathering of some of the key stakeholders to better understand the current landscape and to come up with some concrete actions for the future that the community could embrace and carry forward. Over the next few months and in discussions with colleagues this idea began to take shape and led to "Saving Sounds: Preserving Audio Collections," held in the newly renovated Harry Ransom Humanities Center on the campus of the University of Texas at Austin, July 24–26, 2003.

"Sound Savings" brought together an esteemed group of curators, educators, scholars, and practitioners whose work represents the current thinking in the field of audio preservation. The papers presented here represent the fruit of their individual and collective experiences and, as they reveal, institutions are actively involved in all aspects of audio preservation. But there remains serious work ahead.

We should not be daunted. Many of the preservation issues we face with audio are common to all information entities. Experience is on our side. However, new challenges are magnified by the fact that the quantity of information held in all formats continues to swell. Interwoven with preserving the range and mass of media held in libraries and archives, there emerges the complicated problem of selecting what is to be preserved.

We hope that the information in these papers will provide insight into the current state of audio preservation. As evidenced by the growing interest and activity in audio preservation across the country, clearly the time has come for extended knowledge of the issues, challenges, and former and contemporary solutions employed in the preservation of sound recordings. This summer, "Sound Savings" took the profession to the next step in articulating the most pressing of the challenges we face. Conference attendees—critical stakeholders of the future of audio preservation—articulated seven areas for future action to move the field effectively forward. Our collective challenge in the months ahead will be to develop strategies to attain these goals. Stay tuned.

© 2003 Ellen Cunningham-Kruppa

A Sound Education:
Audio and the Next Great Leap
in Information Studies

Andrew Dillon
Dean, School of Information
The University of Texas at Austin

IT IS SOMETHING OF A TAUTOLOGY to define the field of information studies as a discipline that investigates the properties and behavior of information. But borrowing from Borko (1968) this starting point can be expanded upon to define the field as studying the forces governing the flow of information and the means of supporting optimum access and use. In doing so, we must study the origination, collection, organization, storage, retrieval, interpretation, transmission, transformation, and utilization of information.

As Borko noted, the field of information studies can be seen as both a pure science, developing theories from data on information properties and behaviors, and an applied field, which develops services and products.

While there is much to admire in such a definition, I tend to think of information really as a mix of two components: a representation or product (such as a book, a Web site, an algorithm, a tape, a data set, etc.) and a process of decoding (such as an intelligent reader with a method of access). Considering information as both product and process serves to broaden our perspective of the field and our legitimate areas of enquiry.

The legacy of library and information studies is long and not without honor, but in examining the last century it is clear that in our studies, representation trumps decoding. We have learned about and advocated for storage, preservation, and the development of collections rather more forcefully than we have for sustaining, providing, and protecting the appropriate means of interpretation or decoding.

Rather than talk about the "components" of information so construed, it is perhaps better to consider them as phases, or a natural coupling of structure (or product) and process, where we have too rarely acknowledged the phase of process. To give process its due, we need to extend our studies to human meaning. By this I mean that we need to go beyond emphasizing the artifacts of information (the objects we house and keep) and place them in their proper human context of use. An object without a human is difficult for me to conceive of as information—and while some may place great value on having libraries exist without people, I cannot see the point in us adopting that as a reality (or even an ideal, as some would have it) when the typical citizen spends a large part of her life in the act of interpretation. But the opposite is true, too. We can have no real decoding and interpretation without representations, and the quality of representation is hugely significant to the experiences and advances of our world. Hence my emphasis on regarding information as product and process, combined in a synergistic cycle of use.

Why Sound Matters

So what has this to do with sound preservation? In my view, sound is at the cutting edge of much of what is important now to the field of information studies, and I can see several reasons for this.

▶ Sound is a medium that is intimately tied to tests of copyright limits in our society. And this is most blatantly so since sound is a medium in high demand by consumers.

▶ Music and sound are transcultural in a manner that is not so for text. Whether white men can play the blues may never be resolved in some purists' minds, but there is no doubting that the representations of history and culture that are captured in music can be processed and enjoyed by people outside that culture. The rise of world music, the merging of cultural styles, and the worldwide love of opera by people who cannot speak a word of Italian are testimony to the emotional response people have to music.

▶ The next tidal wave of digital content is rich media, a seamless convergence of audio, video, and text. As yet, true hypermedia of the kind envisaged decades ago has yet to emerge and even the Web, in all its glory, is (with some noticeable exceptions) a text-heavy medium. Audio is the great underutilized resource. Hypermedia in popular use is a visual medium, with audio seen as "extra," but there are signs that this will change.

Spring (1991) noted that at some point in the mid-1980s, a radical shift occurred in computing—without much attention being given to it—when more computing cycles came to be spent on words than numbers.

The question then is, "Are we now continuing that movement from numbers to words to pictures and then sounds?" Perhaps we will end up with them all, and we certainly must if we are to exploit the dream of hypermedia, but to get there we need a far better understanding of sound and its role in information use.

Sound is really an ecological interface to information. By this I mean that sound is a highly refined yet natural source of information for all humans. Among a child's first perceptions are the sounds of his mother's heartbeat. Everyone has favorite pieces of music that evoke strong emotional responses. People buy or consume music in significant numbers. While it is common to talk of library usage rates for books, the ALA's own statistics indicate that about as many people use the library to borrow CDs as they do to use the Internet, the more heralded function claimed for public libraries. Clearly sound has significant status in our lives, but its taken-for-granted nature often causes us to overlook the centrality of audio in everyday use.

Sound at UT's School of Information

It is perhaps something of a cliché to refer to information as the currency of 21st century life but this points to the emphasis now placed on understanding contemporary life through an information-based lens. I am not referring here only to the information economy, important as that is, but to the broader ideals of digital citizenship, information as a right not a luxury, and the need to develop what has often been termed "information literacy" in order to participate fully in today's world.

At the School of Information here in Austin, we consider information as a product and process to be studied and understood across its complete lifecycle, from production to preservation, through management, use, and application.

Preservation and conservation are key components of this information lifecycle, and audio is both a natural element and a complement to other elements in information space. In our curriculum we view sound as belonging everywhere, from Karl Miller's laboratory for sound preservation to the digital media classes where sound is designed into an application to enhance the user experience, again emphasizing the product and process nature of information.

The school is also an intellectual home for the newly announced Knowledge Gateway at The University of Texas at Austin (http://www.gateway.utexas.edu), an ambitious project aimed at providing access for every citizen, via a personalized Internet window, into the resources of our university, including the libraries, collections, museums and much more. Such a project demands audio and ensures that the emphasis on understanding the use and preservation of sound recordings will remain at the forefront of our thinking.

Beyond Sharing: Three Goals for This Meeting

This meeting brought with it a charge. This is the first sound preservation symposium of its kind and our hope is that it will not be the last. However, for this to work participants must move us forward in three directions. I see three goals here:

1. Articulating an agenda
2. Establishing synergies
3. Creating the touchstone

With this gathering of content experts, preservationists, archivists, researchers, teachers, cultural scholars, and foundation representatives there is a real need to find a common voice. It won't be easy but the challenge is important. As a school, we need to know how this discipline is evolving and what our students will need to study in order to participate. Many of us are looking for such an agenda as a major outcome of this meeting.

I have always believed that the future of the information field will be determined largely by how well people from different backgrounds can learn to tackle the problems together. LIS programs have often taken the lead in attracting experts from different fields but the results have not always been as desired. Synergies take effort, first at learning to communicate and then in seeing how combined perspectives and resources can yield better results. I challenge all of you here to find others outside your normal comfort zone with whom to engage. The task then will be to continue this engagement beyond the present symposium.

Finally, this symposium needs to stake the territory. This event should, if we meet the first two goals, require little further effort to be the reference point for others who come later or who could not be here this week. We are in the business of shaping the program and no doubt we will get some of it wrong. But this is the point at which the future of sound preservation should be planned and the point to which others will later refer as the landmark event that started it all. If we are not here for this, why are we here at all?

So, the challenges are immense but so are the rewards. This is a meeting of like minds and the real work now begins. Let's lay the groundwork for progress in this area by engaging in open discussion and sharing lessons learned. The future of sound will depend on us.

References

Borko, H. "Information Science: What is it?" *American Documentation* 19, no.1 (1968): 3–5.

Spring, M. *Electronic Printing and Publishing: The Document Processing Revolution.* New York: Marcel Dekker Inc., 1991.

Review of Audio Collection Preservation Trends and Challenges

Samuel Brylawski
Head, Recorded Sound Section
Motion Picture, Broadcasting, and Recorded Sound Division
Library of Congress

THE SOUND SAVINGS CONFERENCE PRESENTS an opportunity to assess the state of audio preservation programs in the United States in 2003, to examine some issues related to preservation, and to give thought to where we are headed. This is the first national conference dedicated to sound preservation. The significant number of attendees is evidence of increased interest in the challenges of audio preservation and the number of questions we have. Many archives are pursuing the transition in preservation reformatting programs from analog tape as the preservation medium to new digital formats. This transition is probably the single greatest reason that so many people are attending "Sound Savings" and that interest in audio preservation is burgeoning.

For years, sound archivists have been talking about the digital future. No longer do we discuss the digital future—it is the present. Preservationists and archivists are, and should be, conservative and cautious about adopting voguish trends prematurely, especially as we attempt to assure that the audio artifacts of today will be available for study and entertainment centuries from now. Yet, it is clear from my perspective that, in terms of preservation, analog is dead, or at the very least, a dead end.

I will discuss the digital present—and future as well. Yet, the digital revolution, as it is sometimes termed, is not the only subject on our minds, even if many of the other issues relate to the major transition taking place in our archives. I will touch on a variety of related audio preservation issues, but only lightly. Many of them will be covered in greater detail by our speakers this week. While I was asked to look at trends, I find myself thinking as much of the challenges.

Development of Conservation Practices

In managing sound archives we make a distinction between *conservation* and *preservation* of audio materials. Reformatting, that is, conversion of content from one medium to another, is inevitable for most materials in sound archives. But that reformatting, to which we apply the term *preservation,* can be deferred for many years, if not decades, if collections are properly conserved. Unfortunately, no comprehensive, tested, and documented standards exist for the cleaning, storage, and housing of audio collections. Much work is required to develop a set of professional standards based on science for conserving our originals.

We don't know which practices constitute best care for all audio media in our collections. These practices are still being identified. For example, it is now standard practice in archives to provide the best possible care for original recordings, whether they have been re-formatted or not. This wasn't always the case. It wasn't that long ago that masters were routinely destroyed after reformatting. The procedures for conserving audio media are still evolving—more so than for other library media. What is the best packaging for original audio recordings? Do acidic record sleeves contribute to the deterioration of shellac and vinyl discs? Are compact discs potentially harmed by the paper booklets in jewel boxes? Up-to-date published standards for housing and storage of acetate tape do not exist. A number of recently formulated or revised recommendations for storage tapes circulate among archivists verbally but they are not codified within a best-practices manual.[1]

We await research and documentation to help archivists assess the condition of the collections within their responsibility and identify which problems demand immediate attention. All "instantaneous" audio formats (e.g., magnetic tapes and lacquer discs) are known to require eventual reformatting, yet the rates of deterioration for these formats are not known. Given content of equal cultural value and uniqueness, to which medium should an archive give priority for reformatting: hydrolysis-afflicted polyester tapes, audio cassettes comprised of cheap tape stock, or fifty-year-old lacquer discs? Definitive data on which media are most at risk are not known. Archives need documentation to help identify problems affecting media and set priorities for the limited resources available for audio reformatting programs.

In the spring of 2003, the Image Permanence Institute of the Rochester Institute of Technology issued welcome news. They announced that they will undertake a study on the Preservation of Magnetic Tape Collections. It will "focus on the deterioration of magnetic [audio and video] tape and work on creating techniques to help libraries, museums, and archives save their collections."[2]

Digital Preservation

Digital preservation, or reformatting audio onto a digital format, has been discussed for decades and it has been disdained as a viable solution for nearly as long. There have been two major arguments made against digital preservation. The first is that all digital formats are susceptible to deterioration; there is no "permanent" digital format. The other objection has been that common digital formats, such as those employed for compact audio discs, employ algorithms to compress, or reduce, the data required to represent the sound. Compression is usually inappropriate for preservation reformatting because most often the objective is to capture as full and as rich and accurate a reproduction of the original as is possible. With the cost of digital storage diminishing each year, compression is no longer considered to be a necessity.

Indeed, there is no permanent digital format. In the case of analog audio, however, there has never been a permanent, or even long-term, format. Archivists agreed many years ago that 1.5 mil quarter-inch polyester tape on an archival 10-1/2-inch open reel was the best medium for magnetic audio recordings.[3] This format was considered then to be the best available. Yet, all reasonably priced analog audio formats are subject to deterioration and must be copied to new media eventually.[4] Analog quarter-inch magnetic tape stock has become more difficult to obtain as the number of manufacturers diminishes. And, of course, each generation of analog reformatting engenders a loss of content and increase in noise. Sadly, many of the open-reel preservation tapes created in the 1980s have deteriorated faster than the original media whose content they were intended to preserve. The tapes suffer from hydrolysis or "sticky-shed syndrome." The tape binder adhering the recording material to the backing absorbs moisture from the air. Upon playback the tapes squeak and break down. In cases where the original media were saved and conserved, they are often in better condition than the preservation tapes intended to save their contents.

If digital preservation is the new paradigm, what form will it take? A digital format standard comparable to that established by the Association for Recorded Sound Collections in the 1980s has not been established. Many archives are preserving content on recordable compact discs (CD-Rs). This convenient medium plays on existing compact disc players and the best available blank discs cost under $1.00 each. Yet, recordable compact audio discs hold under 700MB of data and are more prone to degradation than manufactured discs. Recordable DVDs hold much more data but at this time there are several competing recordable-DVD formats and these discs are believed to be more fragile than CD-Rs. Given the challenges and limitations of audio preservation on analog tape, it is understandable that archivists have turned to recordable digital media. Judicious archivists see the process as an interim solution, at best, and cover their bets by making multiple copies of the recordable discs and storing the discs in separate locations.

Digital Repositories

Within these proceedings, Carl Fleischhauer describes in detail the Library of Congress's approach to digital audio preservation, the creation of digital *files* intended to be stored in a digital repository. Digital file repositories have been used by banks and the credit industry for decades, and are used by European broadcasting companies for storing files of audio visual content. Digital file repositories are designed to backup data systematically on the preferred storage format of the moment. The data is sustained through any number of shifts in design and configuration of the storage formats. Digital repositories operate on the assumption that there will never be a permanent physical format. Well-designed repositories ensure the persistence of data by validating its integrity periodically when it is copied. The well-planned repository presumes media obsolescence, plans for it, and, according to its supporters, frees the archive community of the futile search for an affordable permanent medium. In the eyes of some archivists, digital repositories are liberating.

Digital repositories will be expensive to build and challenging to operate. They require a sophisticated information technology infrastructure in order to migrate files successfully, keep them accessible, and maintain their integrity. Yet, to managers at the Library of Congress and other large libraries and archives, digital preservation is dependent entirely upon the success of these repositories. Essential in the development of repositories are safeguards to ensure their continued existence in case of a breakdown or catastrophe. Implied in a digital repository is faith in the assurances of a professional IT infrastructure. Repository systems must be networked with built-in redundancy, including mirror sites that can substitute in the advent of an adverse situation. With wise investment of ample resources these complex systems can be built, but only by financially advantaged institutions. Smaller archives must not be left behind. For digital repository systems to be truly successful, they must accommodate collections held by institutions without the resources to build their own systems.[5] With the generous support of the Packard Humanities Institute, the Library of Congress is creating a National Audio-Visual Conservation Center, now under construction in Culpeper, Virginia, 75 miles from Capitol Hill. Concurrent with the planning for this facility is the development of a digital repository for the preservation of audio, video, and other digital collections. Library officials hope that this state-of-the-art storage, processing, and preservation facility will be more than a big, new building. The intention is that it be truly national, perhaps providing storage and repository services for other institutions. The center might also perform duplication services for other institutions if a funding mechanism and a process can be devised which do not compete unfairly with the private sector.

As Carl Fleischhauer points out, digital repositories for audio are not merely collections of sound files. The repository planned by the Library of Congress entails associating *sets* of files to create digital objects. Following the Open Archival Information System (OAIS) model established by NASA,[6] digital objects for sound recordings in the repository will include digital images of record labels or tape boxes and other graphics or accompanying text, in addition to the audio files. The audio files themselves will be very large, recorded at a sampling rates of 96kHz or 192 kHz, with 24-bit word lengths.

The files will be described and controlled administratively by metadata (which can be partitioned into "descriptive," "structural," and "administrative" metadata) about the original recording and its digital files. Structural metadata identifies and organizes the individual files of images and sound that represent a digitized item. The metadata assist with the presentation of these related files from the digital repository. In a repository, structural metadata are called up by program scripts to reconstruct virtually the sound recording's packaging (scanned images of the covers, accompanying text, etc.) and to provide researchers with control over which audio tracks to audition. The Library of Congress is working with other institutions to develop the Metadata Encoding and Transmission Standard (METS) metadata format to facilitate the documentation, maintenance, and presentation of these files.[7] METS is not a universally-accepted standard and we can't guarantee that it will be *the* standard, but I believe that it is the right direction and has the granularity that would facilitate the migration to another standard if necessary.

METS is complicated. Because it requires populating a very large number of fields, at the present, it is time-consuming to create a full METS record. Officials at the Library of Congress hope and expect to "develop tools for automatically creating metadata," as recommended in a study of challenges related to the preservation of digital content.[8]

A preservation manager is quoted by Richard Griscom as stating that, "To date no one can prove that any digital version will survive and be accessible beyond a few decades, despite much talk of migration and emulation. [Libraries should] exploit access capabilities of digital technology and combine them with the longevity of proven preservation methods."[9] This prescription has not yet been disproved, but in the case of audio preservation, unlike print microfilming practices, there is no *proven* analog preservation practice. Digital preservation in some form is here to stay and many capable people are working to ensure that its products remain permanent as well.

Professional Practices

The creation of repositories and files to store in them is but one challenge sound preservationists face. Much work remains to ensure that it is possible to re-format our vast collections. There is an enormous variety of media in need of reformatting. The media fall into broad categories (magnetic tape, discs, etc.), but each medium presents its own challenges for the best possible reformatting. I am not convinced that we are fully prepared to meet these challenges. The development of audio preservation standards and professional practices is in its infancy. Tests must be conducted, best practices proved and documented, and training provided.

Systematic development of a body of professional knowledge about audio preservation will take place on many fronts. The most common preservation practice for tapes afflicted by hydrolysis is to bake them at a low heat in a scientific oven and then re-format them. However, some engineers disdain this approach and argue for alternative methods to "dry" the tapes. I am unaware of documented scientific tests proving any approach and look forward to such research. A catalog of common problems encountered in audio media and their recommended solutions would be of value to all archivists.

Most recognized experts in signal capture from legacy analog media are over sixty years of age, and their methods and tricks may retire with them. There is no systematic program to document senior engineers' most successful procedures and ensure that their wisdom is passed on to future generations. Competence with digital recording tools is not always synonymous with expertise in capturing sound from antique media. Documentation and training for safe and effective capture of sound are as necessary as proven guidelines for housing and storage. A recent study commissioned by the Council on Library and Information Resources concluded that, "Many libraries, and especially smaller ones, need outside help for their preservation programs in the form of advice, instruction, opportunity for learning, contact with those active in the field, involvement in collaborative efforts, and funding."[10]

The Library of Congress National Audio-Visual Conservation Center is intended to support audio-visual preservation education. A training program, administered cooperatively with a local community college, is being considered. Training specialists will have to be funded and hired, and a curriculum written. This program offers great potential but developing it presents significant challenges. In order for such a program to be successful, it will require the support and collaboration of the preservation community.

Other significant work is underway. The conferences of the Association for Recorded Sound Collections (ARSC) have provided important venues for communication among engineers experienced in the playback of legacy media. The *ARSC Journal* has published some of the valuable results of their work together. The association also has taken the lead in developing guidelines for the design of archival cylinder players and stylus design for playback of cylinders.

Playback, or signal capture, methods where nothing physical comes in contact with the recording are being explored by a number of scientists. Vitaliy Fadeyev and Carl Haber at the Lawrence Berkeley National Laboratory are experimenting with using high-energy physics instruments to map the grooves of 78-rpm discs via precision metrology and digital image processing. Their experiments thus far entail both two-dimensional and three-dimensional mapping of the record grooves. With these detailed representations of the forms of the disc grooves, the scientists hope to be able to interpolate corrections to rectify scratches and other groove wear. The approach shows promise but is still in the developmental phase. Currently, it requires over 100 hours to map one side of a 10-inch 78 rpm disc in three dimensions.[11] Independent of this effort, a group from the University of Applied Sciences at Fribourg, Switzerland, and the Fonoteca Nazionale Svizzera are experimenting with taking high resolution photographs of disc recordings, digitizing the photographs, and applying algorithms to extract sound from the images.[12] In a third project, engineers at Syracuse University are developing a laser player for cylinders.

Fadeyev and Haber have expressed hope that their work will lead to large-scale digitization preservation projects. The Library of Congress, too, is investigating the feasibility of mass reformatting. The Library's collections include over 100,000 audio cassettes and 170,000 open-reel tapes. Library managers there suspect that the only hope for preserving the content of many of these recordings is to develop systems to extract the sound without real-time monitoring and adjustments by audio engineers. Several administrators at the Library believe that compromises, or fall-backs from traditional reformatting procedures, will be necessary to assure that audio, such as that on those cassettes, remains accessible for study and enjoyment. Tape playback systems, such as those sold by Quadriga, monitor and transcribe technical metadata, including documentation of tape and signal deficiencies, as tape reformatting takes place. Many preservation specialists see the creation of these devices as positive developments but wish that these devices included more of the tools required for preservation reformatting, such as detailed documentation of and correction for tape variances. The Library of Congress is working with one firm now to create technical specifications for a mass duplication program. These specifications will be well documented and disseminated.

To accomplish mass reformatting more than one source must be duplicated at a time, which will preclude continuous, real-time monitoring of the recorded signal by an engineer. By its nature a mass reformatting system will compromise existing preservation standards. We don't know which compromises will be required, or whether they will be acceptable to archivists and users of audio collections. It will be interesting to observe whether and how these compromises can be agreed upon. What kinds of collections are appropriate for mass reformatting? How will those collections be identified and by whom?

Born Digital and Preservation

In addition to the challenges and opportunities which digitization of sound presents for preservation, this digitization has produced many new methods of distribution of audio which give new responsibilities to archives. In light of these new formats and methods of distribution, many of us are in the process of rethinking what we acquire and subsequently preserve. MP3 file proliferation, and I mean music distributed legally and exclusively as MP3 files, has facilitated more "publishing" of greater amounts of music. Direct marketing over the Internet by musicians challenges archives (especially those with subject or regional focuses) to find new and more thorough ways to identify, collect, and preserve music and other audio from these new, small business sources. Given the ephemeral nature of Web sites, archives will need to act quickly if the content of these sites is to be preserved.

The selection of a format in which to preserve Internet-distributed audio is another question archives must consider. The proliferation of audio on the Web has brought a diminution of the technical quality of much audio. The inherent compression of MP3 files or audio suitable for Web streaming implies a lowering of audio quality standards. If these new distribution models and businesses are unstable, as well as dependent upon compressed audio, might maintaining a collection of high-quality masters be a new responsibility for audio archives?

The present instability, if not disarray, of the music business has other effects upon archives concerned with preservation. Manufacturers no longer claim that compact discs are "permanent." If archives are committed to retaining their content eventually they will have to "rip," or copy the digital discs' content to bit streams. To thwart illicit ripping of CDs some manufacturers are encrypting, copy protecting, and/or watermarking their products. This will make legitimate duplication for preservation more challenging. It has been suggested that another response by the industry to illegal duplication of CDs is to attempt to eliminate compact discs and replace them with mixed-media DVDs or combination DVD/CDs. The hope is to make it more difficult to pirate music and at the same time, make it more appealing in the marketplace. No doubt these DVDs will be more difficult to preserve. The challenge will be to emulate their interactive behavior, in addition to preserving the bit streams.

Collaborative Activism

There is a vast amount of sound that needs to be preserved, and the standards and new efficiencies are not adequate to assure that our audio heritage is secure for posterity. I believe that collaboration among archives is necessary. Given the seemingly perpetual limitation of financial resources available to archives for preservation and the significant number of duplicated holdings among archives, efforts should be made to reduce preservation redundancy as much as possible. The Metadata Encoding and Transmission Standard is intended for more than the management and presentation of files. It has been designed to facilitate sharing of files. Utilization of that potential may be necessary to obtain adequate funding of preservation. It has been observed that to many people, digitization and access are synonymous. Griscom points out that while at one time access used to be secondary to preservation programs, with digitization the driving force is access. Regardless of the legal obstacles to making file sharing between institutions a reality, the distinction between preservation and access has become blurred.[13] Support by the Packard Humanities Institute to the Library of Congress for the creation of the National Audio-Visual Conservation Center has been motivated in part by an objective to improve access to older recordings held by the Library.

Archives should be exploring legal, as well as technical, methods to collaborate on preservation projects and share the products of those projects. This will not be simple. The piracy of recorded music is a great problem faced by the music industry, and convincing intellectual property holders in music to allow archives to share music files will be a significant challenge. In addition, the laws governing recorded sound are complex and often vague. Digital audio reserve systems (the placement of listening assignments on servers to enable convenient and simultaneous use by students) are in common use today. Yet, copyright experts are not in agreement as to whether these are in strictest terms "legal." Since digital repositories result in more than the legally mandated maximum of three preservation copies of a recording, even they may be illegal under a strict interpretation of the law. Recordings of classic radio broadcasts will be particularly difficult to share legally. They are not protected by federal copyright law, but instead by often imprecise state copyright laws and various trade union contracts. As a result, these broadcasts are among the most legally restricted recordings held in archives. Archives will need to work together to establish copyright licenses if they are to share any audio files. Constituencies, both within and outside our institutions, will need to be built, and potential collaborators will have to advocate for legal solutions.

In addition to establishing the legal means to share files, storage and server networks must be established and administered. Congress has charged the Library of Congress with building the National Digital Information Infrastructure and Preservation Program. The NDIIPP was created to help provide the legal and technical blueprint for these networks. The NDIIPP has begun this work by enlisting collaborators and obtaining the counsel of information technology experts, legal authorities, and representatives of content industries. The program will not be the sole province of the Library of Congress. To be successful, it will need to be a national effort that includes participants from the private as well as public sector.

Further assistance will come from the work of the National Recording Preservation Board, created by Congress in late 2000. The board is comprised of 22 individuals, representing archives, the recorded sound and music industries, and relevant professional organizations. It advises the Librarian of Congress on the annual selection of sound recordings to a National Recording Registry of culturally, historically, or aesthetically significant sound recordings. The board, in collaboration with the NDIIPP and the Council on Library and Information Resources, is commissioning studies on a number of audio preservation issues. The recording preservation legislation also directs the Librarian of Congress to "implement a comprehensive national sound recording preservation program, in conjunction with other sound recording archivists, educators and historians, copyright owners, recording industry representatives, and others involved in activities related to sound recording preservation..."[14] The legislation also establishes a National Recording Preservation Foundation, a federally chartered, but independent, corporation to raise funds and award grants for the preservation of audio collections.

We have entered a very promising era for the preservation of audio. There is a broader constituency for preservation than ever before and ever-increasing resources, but of course we need more. We must be careful not to throw money at problems. Many archivists are hopeful that enormous strides will be made in the next ten years: research completed, programs established, and thousands of recordings preserved and made available to the public. This conference is an opportunity to collaborate on the development of an agenda for research and action to address the challenges seriously. I look forward to looking back on this Sound Savings conference as another landmark in a new era of professional sound preservation practice.

Endnotes

1. Two more recent best-practice guidelines are a recommendation to package acetate-based recording tape in boxes which are not airtight, in order to enable necessary off-gassing; and a retraction of the directive to "exercise" (slow-wind) tapes periodically. The latter is no longer a recommended practice. Unfortunately, word of the change in recommended practice has not been adequately circulated to archivists in charge of audio tape collections .

2. "RIT Studies Increasing Shelf Life for History Stored on Tape." Rochester Institute of Technology press release, June 25, 2003.

3. Association for Recorded Sound Collections, Associated Audio Archives, *Audio Preservation: A Planning Study: Final Performance Report.* (Silver Spring, MD: Association for Recorded Sound Collections, 1988).

4. Pressed vinyl discs have been proposed as a preservation medium but vinyl, too, degrades eventually, and the cost of creating master discs and pressings for every hour of audio to be saved would be prohibitively expensive. The Church of Scientology commissions platinum analog discs of recordings of founder L. Ron Hubbard and these may well be a permanent medium. But if such a solution is viable, it is so only for a limited body of work as each disc must cost the Church thousands of dollars to produce.

5. Committee on an Information Technology Strategy for the Library of Congress, Computer Science and Telecommunications Board, Commission on Physical Sciences, Mathematics, and Applications, National Research Council. 2001. *LC21: A Digital Strategy for the Library of Congress* . (Washington, D.C.: National Academy of Sciences).

6. ISO Archiving Standards—Reference Model Papers. http://ssdoo.gsfc.nasa .gov/ nost/isoas/ref_model.html

7. Metadata Encoding and Transmission Standard. http://www.loc.gov/standards/ mets/

8. National Academy of Sciences. *LC21.*

9. Richard Griscom, "Distant Music: Delivering Audio Over the Internet," *Notes* (March 2003).

10. Anne R. Kenney and Deirdre C. Stam, *The State of Preservation Programs in American College and Research Libraries: Building a Common Understanding and Action Agenda.* (Washington, D.C.: Council on Library and Information Resources, 2002). http://www.clir.org/pubs/reports/pub111/introsum.html

11. "Sound Reproduction R & D Home Page." http://www-cdf.lbl.gov/~av/

12. Ottar Johnson, Frédéric Bapst, etc., "VisualAudio: An Optical Technique to Save the Sound of Phonographic Records." *IASA Journal* 21 (July 2003): 38–47.

13. Griscom, *Notes* (March 2003).

14. *National Recording Preservation Act of 2000,* Public Law 104-474.

Pictorial Guide to Sound Recording Media

Sarah Stauderman
Preservation Manager
Smithsonian Institution Archives

THIS IS A TEXT VERSION of a Web site (http://www.video-id.com/) created to illustrate the many prominent audio formats that may be found in libraries, archives, museums, and other collecting institutions. It is meant as a resource for conservators, curators, collections managers, and others who need to know the formats and types of audio materials in their collections. Since the first attempt to capture sound in a solid medium there has been a proliferation of media types and formats.

On the Web site it is possible to view thumbnail images of the prominent audio formats and read a short essay on the formats for the particular time period. Because preservation is an important concern, additional information about the materials that make up the formats is provided. The article that follows is an adaptation of the Web site developed in 2003 with the help of Paul Messier, Boston Art Conservation.

CYLINDERS

History

The first sound recordings were made with foil covered brass cylinders (1877–79, Edison) which came to be known as tinfoil records. These impermanent recordings were eventually replaced with wax- or plastic-based cylinders of varying dimensions that could be either prerecorded or recordable depending on the formulation and manufacture. The height of wax and plastic cylinders' popularity is from about 1887 (Bell-Tainter/American Graphophone Co.) to 1929 when the Edison Company discontinued its commercially recorded cylinder product. However, cylinder recorders were used to a great extent in live recording of ethnographic field notes as well as for office dictation, so archival collections may have cylinders dating from the 1930s through the early 1960s.

The length of sound recordings on cylinders depend on the dimensions of the cylinder, the numbers of grooves per inch, and the rotations per minute (rpm). Soft wax cylinders (cylinders with 100 grooves per inch) ran approximately 2 to 2.5 minutes of playing time. "Longer Play" cylinders with 200 grooves per inch ran twice as long, up to 4.5 minutes. Cylinders also have different rotations per minute, depending on the manufacturer and advances in technology, such as 120 rpm, 144 rpm, or 160 rpm.

Many manufacturers produced cylinders in their heyday, but the only substantial difference in the recording or playback of different cylinders corresponds to the diameter of the cylinder, which requires a corresponding size armature to hold it. Cylinders were recorded acoustically (also known as mechanical recording). Acoustic recording is defined as sound waves affecting a diaphragm attached to a stylus that will impress a sound track (corresponding to the sound waves) onto a recording medium.

Materials

Soft Wax Cylinders (1887): Wax cylinders were the first of the cylinders and were usually direct original recordings, though some prerecorded soft wax cylinders exist. In the first few years of their manufacture and use, they were an ivory or cream color but in later years were a medium brown color. On occasion they were used solely for dictation, and the wax could be scraped off to present a new surface for recording. Wax cylinders were made of various waxes, resins, soaps, and oils with additions of colorants, anti-fungal oils, plasticizers or lubricants, and hardeners. Wax cylinders could be solid or could have a cardboard core. Ward (*A Manual of Sound Archives Administration*) refers to two recipes (p. 125) for making these cylinders:

A typical recipe for the composition from which brown wax cylinder blanks were moulded was 12 lb. stearic acid/1 lb. caustic soda/1 lb. ceresin or paraffin wax/1 oz. aluminum oxide. Other ingredients used in Edison wax cylinders were "burgundy pitch," frankincense, colphony, spermaceti, and aluminum stearate.

Molded Cylinders (1902–03): Prerecorded cylinders became available, made of hardened wax or metallic soap (this also provided a sharper, superior sound). These were fragile and brittle. Cellulose nitrate cylinders with cardboard or plaster cores became available after 1908 and culminated in the "Blue Amberol" so-called indestructible cylinder in 1912.

Sizes (Diameter/length)

Cylinders of different diameter cannot be played on the same cylinder machine. Here is a sampling of sizes for cylinders: 1 5/16" diameter x 4" length; 2 1/4" diameter x 8" length; 3 3/4" diameter by 6" length; 5" diameter x 4" length.

Manufacturers

Edison Phonograph Works, London Stereoscope Co., North American Phonograph Co./Jesse Lippincott, The Columbia Phonograph Co., The American Graphophone Co./Bell-Tainter/Volta Graphophone Co., American Talking Machine Co., Pathe-Freres, and Edison-Bell Consolidated Ltd.

DISCS

History

The grooved disc (platter, record) was an invention of Emile Berliner in 1887. Advances over the next 75 years created dozens of sizes (diameters) and colors of discs, and a variety of rotation speeds (beginning at around 70 rpm) depending on the manufacturer and materials.

Discs are made through one of two processes. In the master and mother process, a recording blank is etched to create the matrix for a permanent mold or stamping for prerecorded discs. In the instantaneous process, a stylus cuts a groove in a blank disc to capture original recordings. Discs are usually cut laterally (the groove has side-to-side impressions), though for a time in the early 20th century, they sometimes were cut vertically (so-called "hill-and-dale" impressions), depending on the manufacturer. Disc recordings span the acoustic and electric method of recording. Many discs, especially instantaneous discs, are recorded inside-out.

In general there are three eras of disc materials found in collecting institutions:

▶ shellac-type recordings (1897–c.1948)
▶ instantaneous discs made either of aluminum or cellulose nitrate on a core (cellulose nitrate discs are also known as acetates and lacquers) (1930s–1940s)
▶ thermoplastic discs of polyvinyl chloride or polystyrene (LPs and microgroove discs) (1948–present)

There are a variety of master recording blank materials as well as unusual discs developed for specific markets. The chronology of discs, including material characteristics, diameters, and rotations, is as follows:

1887 | Berliner disc
[matrix] glass covered with lampblack; traced and lacquered; photoengraved

1888 | Wax recording blank
[matrix] zinc disc coated with wax; traced; etched with acid

1888–97 | Berliner record
[prerecorded] hardened latex (latex rubber vulcanized with sulfur) also known as vulcanite; easily malformed

c.1888–97 | Celluloid disc
[prerecorded] cellulose nitrate; brittle

1896 | Solid wax disc
[matrix] solid wax disc

1897–c.1948 | Shellac disc
[prerecorded] clay (Byritis), powdered shellac, lampblack, cotton fibers (originally known as Durinoid); formulations change from brand to brand, and other resins, plasticizers, hardeners, and fillers may be added. 7 inch, 10 inch, 12 inch, 16 inch. 70 rpm, 78 rpm, 30 rpm

1906 | Laminated shellac disc
like the shellac disc but with a core of heavy cardboard

1930s | Aluminum disc
[matrix or instantaneous] aluminum. 12 inches or smaller

1934 | "Acetates" or "lacquers" or "direct-cut discs"
[matrix or instantaneous] cellulose nitrate lacquer on an aluminum, glass, or zinc core; or cellulose acetate on a core. 10 inch, 12 inch, 13 inch, 16 inch (also unusual sizes larger and smaller)

Early 1940s | Dictation discs
[instantaneous] "plastic" discs under the names Voicewriter, Gray Manufacturing, Audograph

1948 | Long-playing (LP) disc
[prerecorded] polyvinyl chloride with stabilizers. 10 inch (1948–1960s); 12 inch (1948–present) 33 1/3 rpm

1948 or 1950 | 7" Microgroove disc or "45s" | [prerecorded] polystyrene or polyvinyl chloride with stabilizers. 7 inches. 45 rpm

MAGNETIC MEDIA

History

Though magnetic recording proved viable and available as early as 1898 through the wire recording inventions of Poulsen, it was not until the advent of magnetic tape in the 1940s that magnetic media became popular. In part, the reason for the delay in using magnetic wire recordings was that the technology produced relatively inferior playback fidelity. Improvements in recording and playback technology coincided with the rise in the technology to produce magnetic tape. The first magnetic tape was perfected in Germany in the 1930s and during the WWII years; Allied Forces captured samples of tapes and tape machines at the end of the war and brought them to Britain and United States for development. By the late 1940s, Ampex and EMI had developed broadcast quality audio reel-to-reel tape. The Sony Walkman (portable cassette player), introduced in 1981, made the 1980s the decade of the [compact] cassette, although the cassette had been available since the 1960s. Magnetic recording has spanned the acoustic, electronic, and digital recording age.

Formats and Tape Track Configuration

Formats, both analog and digital, can usually be identified by the shape or size of the tape cassette or reel. In addition to Format configurations (tapes that will only play back on the machine they were built for), there are Tape Track Configurations or Layouts. Unless a written record has been made about how the recording was made, it is difficult to distinguish the different layouts, and it is possible that important sound information can be lost in reformatting. It is essential that the playback head is the same configuration as the track to optimize playback.

Full Track (monaural)

One track, one channel; typically on a 1/4" reel-to-reel tape but can pertain to any width (special purpose 1/2", 1", 2"). Usually left smoothly wound "tails out" (backwards) in professional applications and environments. Note: an obsolete special purpose full track format used in TV and film applications (for synchronization) superimposed a two-track, (push-pull) 60-cycle signal across the entire width of the tape.

Half Track

Also known as two track monaural = two tracks recorded in opposite directions, one channel each; 1/4" reel-to-reel and monaural cassette

Twin Track, also known as "two-track" or "two-track stereo"

Two tracks going in the same direction, each track is a channel; typically on a ¼"

reel-to-reel. But can also be used as a half-track mono. This also applies to ¼"
in broadcast cartridges that outwardly resemble the old 8-track consumer cartridges. If "stereo" (not used as mono half-track), these are usually left smoothly
wound "tails out" (backwards) in professional applications and environments.
Different professional version manufacturers have/had different track widths.
Note: There's a special professional two-track 1/4" format that contains "CTC":
center track time code. It's a third channel containing speed & timing information for synchronization to video tape and film.

Quarter Track, also known as four-track stereo

Four tracks alternating directions: the 1st and 3rd track comprise "SIDE A" and
the 2nd and 4th comprise "SIDE B." Tape stock is 1/4" wide. Note: channel 1 =
side A left; channel 2 = side B left; channel 3 = side A right; channel 4 = side B
right.

Four-track, also known as four-track quad

Four tracks, each going in the same direction, each comprising their own channel: 1/4" and 1/2" reel-to-reel. These should be left smoothly wound "tails out"
(backwards) as per professional applications and environments.

Stereo cassette (Phillips, aka standard format or compact cassette)

Four tracks, the 1st and 2nd track are "SIDE A"' and the 3rd and 4th track comprise "SIDE B." The standard speed is 1/78 ips but some recorders optionally
recorded and played back at 3.75 for better fidelity. Tape stock is 1/8" wide. In
addition to enabling or preventing recording, the tapedeck senses the presence
of holes on sides of the cassette's case to properly accommodate the type of tape:
"type I," "type II," or "type IV."

Eight-track stereo cartridge

Eight tracks, each in its own channel, going in the same direction, making 4
sides; tape stock is 1/4" wide.

16-track

Found on 1" and 2" reel-to-reel.

24-track

Found on 2" reel-to-reel.

Materials

Wire recordings: stainless steel wire; some mid-1920s are on 6mm "wire tape."

Tape: Analog and digital tape are composed of a base, binder, and pigment. The base can be paper (c. 1946), polyvinylchloride or PVC (1946–c.1950), cellulose acetate (1946–mid-1950s), or, most commonly, polyester (mid-1950s). PVC tapes frequently do not have a binder, but otherwise the binder is polyurethane. The pigment can be, most commonly, ferric oxide. Other pigments include chromium dioxide, metal particle, and metal evaporated tape.

Formats

(1898) 1945–55
: Wire, wound on metal or plastic spools

1944–present
: Reel-to-reel tape; in 1/4", 1/2", 1", and 2" widths and multiple diameters wound on metal or plastic spools.

circa 1950s–90s
: Continuous loop cartridges for automatic playback, major names include Mackenzie Program Repeater and Telex "Cart"

1958
: RCA Cartridge "Stereo Tape"

1962
: Muntz Stereo Pak 4-track

1963
: Philips Compact Cassette and its miniaturized version, the microcassette (late 1960s), 1/8" wide tape, runs at 1 7/8 inches per second (ips).

1965–80
: 8-track Tape; a cassette-type format.

1966
: Playtape, 2 track

1970s–present
: Use of videotape cassettes for mastering original recordings (see VHS, Betacam, 3/4" Umatic) using conversion kits (http://www.video-id.com)

1977

Elcaset, 1/2" tape, runs at 3 3/4 ips

1988

Digital Audio Tape or R-Dat

1991

MiniDisk (Sony), this is a crossover format with properties of magnetic and optical recordings (see the optical disk page)

1992–96

Digital Compact Cassettes (Philips)

1990s

Data Cartridges for storing compressed music files such as MP3, .wav, or other digitized sound

Stability

Wire is stable but it is an obsolete format; its primary problems are mechanical. For instance, it easily tangles or breaks. In all studies of magnetic tape media, the least stable part is the polyurethane binder regardless of whether it is an analog or digital recording. The life expectancy of magnetic media is 10 to 30 years according to studies by the National Media Laboratory and the Council of Library and Information Resources.

OPTICAL MEDIA

History

Light (hence, "optical")—typically laser, but sometimes polarized—is used to write and read the data encoded on the recording surface of these media. In 1983 Sony introduced the Compact Disc (CD) and in 1996/97 the DVD was introduced. The magneto-optical disc (MO Disc), introduced in 1992 as the MiniDisc for audio files, is a hybrid of magnetic and optical technology; it was intended to replace the compact cassette and supplement the CD. Instead, the technology that allows consumers to record files in MP3 data files (the personal computer, MP3 players, and the Ipod) seem to have replaced the compact cassette and MiniDisc, while CDs are slowly being replaced by DVD-R.

CDs and DVDs have several different formats (see below). The MiniDisc is one format of the magneto-optical disc technology. The MiniDisc system was introduced in the consumer audio market as a new digital audio playback and recording system. Magneto-optical disc recording technology has been used for computer data storage systems since the mid-1980s. The principal of MiniDisc is based on the Curie Temperature of magnets. In essence, at the Curie Temperature (usually, for MiniDisc, at 200 C) a magnet will lose its magnetic field and can be reoriented. In magnetic recording systems, currents induced in magnetic heads create and read data; in magneto-optical systems, laser light writes data, and polarized light reads data from the disk because the light is reflected differently depending on the magnetization of the substrate (this is known as the Kerr effect).

Media

Magneto-Optical Disc: Known as the "MiniDisc" for recording sound files, this is a thin (5mm) magnetic film disc, 63 mm in diameter, enclosed in a hard square case. Very high heat can eliminate the magnetic flux that encodes the information. Otherwise, there is very little information on the longevity of this medium.

CD: The CD is a laminated disc of polycarbonate plastic, a reflective metal (aluminum, gold), and lacquer. There are two physical sizes: 12 cm (4.7 inches) and 8 cm (3.1 inches), both 1.2 mm thick, made of two 0.6 mm substrates glued together.

DVD: Digital Versatile Disk or Digital Video Disk has the same dimensions as a CD. A DVD is essentially two thin CDs laminated together and contains additional adhesives and temperature sensitive dyes. A DVD disc can be single-sided or double-sided. Each side can have one or two layers of data. The amount of video a disc can hold depends on how much audio accompanies it and how heavily the video and audio are compressed. There are many formats of DVD, depending on the ability of the DVD to be written once or multiple times, and whether it can be erased.

In both CDs and DVDs, damage caused to discs comes from poor storage and handling, although there may be inherent vice in the materials used to create the disc. Unfortunately, it is almost impossible to characterize the materials of these discs because manufacturers change materials frequently. As with all machine-readable formats, the ongoing development of technologies may render a CD or DVD obsolete (unreadable) even when the medium is stable.

Formats

MiniDisc

Is one format of the magneto-optical disc storage technology.

CD

There are multiple CD formats but they look the same. Different formats may require different spin speeds. Earlier versions of CD-R may require slow spin speeds.

- CD-Audio (or audio CD) 1982 to present: Most commercially recorded music is sold on audio CDs that contain approximately 60 minutes of audio data; information about how to play back the CD-Audio is encrypted in the disc itself so that the machine that plays can be a "dummy." Audio CD players cannot play CD-Rs, for instance, unless the CD-R carries information about playback. CD-Audio records material at 16 bits and 44.1 kHz sampling rate. CD-Digital Audio is the same as CD-Audio.
- CD+G (Compact Disc plus Graphics): used for graphic and audio applications such as Karaoke. Requires a special reader.
- CD-I (Compact Disc-Interactive): used for videogames that include music and graphics. Runs on a player that attaches to a monitor or television or used for exhibition kiosks.
- CD-R (Compact Disc-Recordable): a version of CD on which data can be recorded but not erased. CD-ROM (Compact Disc-Read Only Memory): Like CD-Audio, but it allows computer data to be stored; therefore, software programs and other computer-based applications typically come on a CD-ROM. Typically runs on a computer.
- CD-RW (Compact Disc-Rewritable): A version of CD on which data can be recorded and erased and re-recorded in the same physical location of the disc.
- Video CD (V-CD): A standard for displaying full motion pictures with associated audio on CD. The video and sound are compressed together using the MPEG standard and recorded onto a CD Bridge disc.

DVD

There are multiple DVD formats. Again they are indistinguishable from a CD or from other DVD formats.

- DVD-Audio: An audio-only storage format similar to CD-Audio. DVD-Audio differs, however, in offering 16-, 24-, and 24-bit samples at a variety of sample rates from 44.1 to 192 kHz. DVD-Audio has at least a double fidelity of the standard CD, and can also contain video, graphics, and other information.
- DVD-R (DVD-Recordable or DVD minus R): A version of DVD on which data can be recorded but not erased by a disc drive. It has a capacity of 4.7 gigabytes. There are three versions of DVD-R: DVD-R(A) *recordable for authoring*; DVD-R(G) *recordable for general use*; and DVD+R *DVD plus recordable*
- DVD-RAM (DVD-Random Access Memory): A rewritable DVD.
- DVD-ROM (DVD-Read Only Memory): A data storage disc. Will not play in DVD-Video players (for movies); however, DVD-ROM drives will play DVD-video if the proper software is installed.

- DVD-RW (DVD-ReWritable or DVD minus RW): A data storage disc that can be written to approximately 1000 times. Compatible with most DVD-Video and DVD-ROM drives.
- DVD+RW (DVD plus RW, or DVD plus ReWritable): A data storage disc for all content. Compatible with most DVD-Video and DVD-ROM drives.
- DVD Video: Used for viewing movies and other visual entertainment using high-quality MPEG2 or MPEG4 video and igital surround sound. The total capacity is 17 GB if two layers are used on both sides of the disk.

BIBLIOGRAPHY AND RESOURCES

Associations

AES Audio Engineering Society http://www.aes.org/

ARSC Association of Recorded Sound Collections http://www.arsc-audio.org/

Books

Butterworth, W.E. *Hi-fi: From Edison's Phonograph to Quadrophonic Sound*. New York: Four Winds Press, 1977.

Byers, Fred R. *Care and Handling of CDs and DVDs*. Washington, D.C.: Council on Library and Information Resources and National Institute of Standards and Technology, 2003.

Dale, Robin, Janet Gertz, Richard Peek, and Mark Roosa. *Audio Preservation: A Selective Annotated Bibliography and Brief Summary of Current Practices*. Chicago: American Library Association, 1998.

McWilliams, J. *The Preservation and Restoration of Sound Recordings*. Nashville, TN: American Association for State and Local History, 1979.

Pickett, A.G. and M.M Lemcoe. *Preservation and Storage of Sound Recordings*. Washington, D.C.: Library of Congress, 1959.

Read, O. and W. L. Welch. *From Tin Foil to Stereo: Evolution of the Phonograph*. Indianapolis, IN: Howard W. Sams & Co. 1959.

Van Bogart, J. *Magnetic Tape Storage and Handling: A Guide for Libraries and Archives*. Washington, D.C.: National Media Laboratory and Council on Library and Information Resources, 1995.

Van Praag, P. *Evolution of the Audio Recorder*. Waukesha, WI: EC Designs, Inc., 1997.

Ward, A. *A Manual of Sound Archive Administration*. Hants, England: Gower Publishing, 1990.

Web Sites on History and Media Types

[listed here with permission; current as of 9/2003]

The History of Recorded Sound Technology http://www.recording-history.org/HTML/start.htm

Optical Storage Technology Association http://www.osta.org/

Recording Technology History http://history.acusd.edu/gen/recording/notes.html

The Edison Museum http://www.edisonnj.org/menlopark/birthplace/northamericanphonograph.asp

Consumer Audio http://mhintze.tripod.com/audio/default.htm

Edison Cylinders http://www.nps.gov/edis/pr_loc_rec_020103.htm

The Internet Museum of Flexi / Cardboard / Oddity http://www.wfmu.org/MACrec/index.html

DVD FAQ http://www.dvddemystified.com/dvdfaq.html

Web Sites on Preservation

Library of Congress Preservation of Sound Recordings FAQ http://lcweb.loc.gov/preserv/care/record.html

Audio Preservation Bibliography and Web Reference http://palimpsest.stanford.edu/bytopic/audio/

Vendors

The Cutting Corporation http://www.cuttingarchives.com/head/faq.html

VidiPax http://www.VidiPax.com

Art Shifrin http://www.Shifrin.net/

Steve Smolian http://www.soundsaver.com

Richard Hess http://www.richardhess.com

Acknowledgments

The author wishes to thank Paul Messier, Boston Art Conservation, for formatting the images and text for the Web site; Art Shifrin and Richard Hess for clarification on format types and fascinating details about them; and Sam Brylawski, Allan Goodman, and Larry Miller, Library of Congress, Motion Picture Broadcast and Recorded Sound Division, for information and allowing me to photograph their collection of formats.

Surveying Sound Recording Collections

Hannah Frost
Media Preservation Librarian
Stanford University Libraries

THE COUNCIL ON LIBRARY AND INFORMATION RESOURCES (CLIR) is embarking on a new survey project concerning audio materials in academic libraries. The project "aims to guide decision makers and managers, not staff closer to the ground, about what problems exist and how they can be ameliorated" (Smith 2003). CLIR has taken a number of steps to prepare for this effort. For example, it commissioned the preparation of a report titled *State of Recorded Sound Survey of Surveys*, which concisely summarizes the aims of ten independent surveys conducted between 1995 and the present, and considers the results of these surveys together, from which a number of conclusions are drawn:

▸ Budgets for the management of recorded sound collections are limited or non-existent.

▸ Many collections managers may be in need of education concerning the differences between master copies and listening copies, media storage requirements, and the urgent need for preservation of certain audio formats.

▸ Proper storage conditions for recorded sound are understood but not always followed.

▸ The most widely held sound recording format in collections appears to be magnetic audio cassettes . . . [a format] not recommended for long-term storage.

▸ While some formats of recorded sound media are in greater decay and more urgently in need of preservation than others, all formats are in some need of preservation, identification, and cataloging.

▸ There is no authoritative dataset describing the content, location, and preservation status of the nation's inventory of recorded sound held in special and private collections. Furthermore, there appears to be no single approach to gathering such data (Allen 2003).

To librarians and archivists with experience working with audio materials, these statements will not sound off the mark.

Many of the factors inhibiting the use and preservation of sound recordings in libraries and archives are symptomatic of a single fundamental problem: the lack of documentation and understanding about the sound recordings, their contents, and their condition. It seems that research libraries and archives simply do not know what they have or where to focus their attention, and this has been the case for far too long. One effective way to address this situation is to conduct a collection survey and today there are a host of good reasons to do just that.

Why Survey Audio Collections?

A number of indications suggest that the coming decade will be for audio materials what digital imaging was for photographic and other printed materials in libraries and archives during the 1990s. Foremost among these is digital audio technology, which is the inevitable, acceptable, and feasible choice for preservation reformatting of audio information. The time has come to abandon the wait for an analog "archival medium" or "format" for the long-term storage of media materials such as sound recordings and to prepare for migrations to the digital domain. The ubiquity of digital audio technology has fomented widespread interest in access to media content, both among scholars and the general public alike. As a result, the number of potential users of sound recordings in collections is on the rise. With this newfound, popular appeal, our collections can enjoy a new life among a new audience. Of course digital audio technology has also brought to the fore sticky intellectual property rights issues, but these issues in no way prevent cultural institutions from conducting honest preservation activities like collection surveys. Indeed, survey results may catalyze action on the part of rights holders to collaborate with archives and libraries to find balanced solutions that enable enhanced access to sound recordings in collections.

Meanwhile, it seems that within the research library community there is a greater awareness of the problem of technology and format obsolescence. This awareness may be due to the tremendous threat that obsolescence now poses to the longevity of digital information. We may come to find that a wider understanding of the relationship between obsolescence and the preservation of digital materials brings about a wider understanding of the hindering role obsolescence has always played in the preservation of sound recordings and other technology-driven materials in collections. Other indications that audio preservation is coming of age are the newfound sources of financial support. With the formation of the National Recording Preservation Board, federal grants for audio preservation will be available soon.

Over time, these grants will be increasingly competitive, so those who are prepared early to embark on audio preservation projects will be in the best position to take advantage of the grants.

To make the most of this unprecedented opportunity to preserve and provide access to sound, collection caretakers need to be able to make intelligent decisions. But effective, responsible decision making requires accurate and thorough information about the collections and their condition—precisely the information not readily available to decision makers. This paper describes how to end this perpetual state that is hindering the preservation of audio materials by conducting a survey. Topics addressed include: what a survey is and what it can accomplish; how an audio survey differs from a survey of print materials; factors and considerations in survey planning; and the conditions and traits to observe and document during the survey process. Sampling methodologies, as well as the design and application of the data collection instrument, are briefly discussed. The aim of the paper is to demonstrate that a collection caretaker does not need to be an audio expert in order to get a feel for the condition of audio recordings in the collections under her care, for once she explores the materials and becomes familiar with the range of intrinsic and potential problems, it is possible to develop a sensibility for their relative condition and thus initiate an effective audio preservation program.

What is a Survey?

Survey work is research. It is a formal process of gathering and analyzing data about a given population, a collection of things. It is science, more than it is art, though as the work progresses, the artful aspects of survey work emerge. A survey is an opportunity to study items in a way that day-to-day routine work does not allow. And while often a collection survey involves random sampling, the process is anything but arbitrary. Make time for it and take it seriously. A survey will be most effective, however, if it is conceived as not simply a process of observation, data collection, and analysis. A survey forms the basis of a plan of action. As Barbara Appelbaum and Paul Himmelstein keenly noted, "For a survey to be useful, it must be part of a wider plan for collections management" (Appelbaum and Himmelstein 1986). A collection survey becomes a part of the collection's history, the data gathered serving as a reflection of the collection's past, present, and future. The institution must be committed to taking action based on results of survey work.

With careful design and implementation, a survey can lead to a number of productive outcomes: the character and extent of holdings are defined; the condition of items, housings, the storage environment, and disaster plans are assessed; and reformatting priorities are established as the deterioration or damage exhibited by collection items is observed. Also, the data compiled in a survey can be employed to estimate the rate of future growth of sound recording media within incoming collections (Seubert 2003), as well as to project costs for reformatting and rehousing. Furthermore, information pertaining to the contents of recordings that supplement existing catalog records may be readily gathered in the course of a survey. A survey may ultimately reveal opportunities to collaborate with peer institutions with like holdings in coordinated preservation reformatting projects in order to leverage limited funding effectively. Finally, with the information generated by a survey, an institution can contribute its knowledge of its collections to those broad-reaching efforts, like the one spearheaded by CLIR, to arrive at a meaningful understanding of the state of the nation's recorded sound heritage.

How is an Audio Survey Different?

Many collecting institutions have experience in conducting surveys of their holdings. In research libraries and archives, past surveys overwhelmingly have been focused on paper-based materials. Items in non-paper based formats have not enjoyed the same level of attention for a number of reasons. One primary reason is that, to many curators, archivists, and librarians working with broad collections in diverse formats, the preservation issues of sound recordings (not to mention video recordings) are shrouded in arcane jargon and technical complexity. Only in recent years has preservation education broadened its coverage to address media formats in depth. Collection caretakers interested in turning their attention to audio may find that they need more upfront training, especially in areas of audio format identification and signs of deterioration, before they are ready to embark on an audio survey.

Another distinguishing feature of a survey focused on audio materials is that, unlike a book or manuscript, whose content is overt and can be assessed for its legibility and research value by simply looking at it, a sound recording must be played back in order to judiciously assess its contents and the quality of the recording. Unfortunately, for many institutions, listening to recordings as part of the survey process may not be possible or practical due to time constraints or a lack of available equipment. Indeed, the fragile condition of many recordings in archives is such that they should only be played back by an expert with experience in handling original recordings and operating the equipment necessary to access them. This situation forces the surveyor to recognize and accept the risks inherent to conducting an audio survey that does not include a playback component. If selection decisions are based solely on visual inspection, the potential exists to either pass over an important artifact if it appears to be in good condition, failing to identify and address its hidden needs, or alternatively, to expend attention and resources in preserving an item deemed to be in bad condition, only to discover later after hearing the recording, that its condition was not so dire and its contents are of little historic interest.

Selecting audio materials for preservation reformatting with little knowledge of the recording itself can have other unpredictable consequences. For instance, in the case of a digitization project, if you do not know the temporal duration of the recordings, it is difficult to estimate the number of files (and their size in bytes) that will be produced in the course of the project. Tapes in archival collections can often exceed two hours in duration; some hold as many as four hours of sound. As digital files of these originals accumulate, so do file storage costs, and file management becomes increasingly complex. Another concern is lack of knowledge about the quality of the recording. It is frustrating to spend precious preservation resources on reformatting a recording, only to hear that the quality of the original is so poor or the sound is unbearable to listen to; more time and resources may be necessary to "de-noise" and boost the recording in order to produce an access copy someone can use effectively. If one is flexible, open, and attentive, vagaries such as these add a sense of adventure and discovery, rather than frustration, to the work of sound preservation. In any case, the results of the CLIR Survey of Surveys serves as a reminder that most audio materials require some degree of preservation attention, whether due to the uniqueness of many recordings in archives, the likelihood of media instability and deterioration, or, at some point in the future, the inevitable lack of access to appropriate playback equipment. Thus, more often than not, preservation efforts and resources effectively directed at audio materials are, by and large, efforts not wasted. Furthermore, in fact, much can be learned about the state of an audio recording, and decisions about whole collections can be made simply based on the process of observation. The trick is knowing what to look for and being prepared to record the observations.

Survey Planning and Design

In planning an audio survey, it is crucial to define its purpose. Articulate in writing the questions to be answered by the survey. Be clear about the intended outcome of the survey, what actions may be taken as a result of data gathering and analysis. Delineate the scope of the survey. Is it a collection-level or an item-level survey? Is the survey limited to certain formats, certain parts of the institutions' holdings, or will it be comprehensive? Will existing preservation masters or listening copies be evaluated, or only original materials? Define the duration of the survey and be sure to limit it to a reasonable period of time. A survey with seemingly no end in sight is subject to drift, likely never to be completed or to accomplish its intended goals in full. Finally, consider the survey methodology and how it serves the underlying goals of the project, being sure to keep careful documentation of the procedures for future reference.

The survey design must take into account a number of factors. How much information about the audio materials exists? Do the items have catalog records or are they only represented in listings on paper? Do existing records provide complete and accurate information about the collection's size, location, and the formats it encompasses? How are the materials organized and stored? Are audio recordings dispersed among discrete collections with items in other formats or are they stored together by format? How much data entry is the survey staff willing and able to do? Consult with catalogers, curators, and archivists to ask questions about finding aids, the history of the collections, past storage conditions, and any available documentation about the items to be surveyed, such as acquisition files and processing records, for it may contain crucial clues that save time and influence the survey approach. Consideration of these factors will determine if a full assessment of each sound recording is necessary and feasible. If the size of the collection is manageable given the available resources, then a full assessment is recommended. However, for most institutions, a full-scale, item-by-item survey is neither affordable nor practical. In these cases, a sample survey is called for.

Sample surveying is a powerful tool. It is remarkable that one can gather representative data (with a confidence interval of 95% ±5%) about a collection encompassing 50,000 items by surveying only 381 items (Powell 1991). But the choice and application of a sampling methodology is critical to the success of a survey. If a sample is not selected carefully, then the data collected in the survey will not accurately portray the traits of the larger collection it is intended to represent. There are several sampling methodologies used in survey research. Typically in surveying collection materials, one of four methods is employed: systematic sampling, judgment sampling, simple random sampling, and stratified sampling. The decision about which sampling methodology is appropriate depends on how a given collection is organized or arranged, how heterogeneous or homogenous it is, and which variables are to be observed. The first two methods, systematic and judgment sampling, are both attractive for their ease of implementation; these are non-random methods, and as such, are inherently prone to bias or systematic error. On the other hand, simple random sampling and stratified sampling are methods that, if applied correctly, are less likely to introduce bias into the survey results, yet they are not immune from random error, also known as noise. There always remains the chance of encountering noise in the course of surveying. In the face of this fact, the surveyor must ask from the outset: "What are the consequences of being wrong? When it comes time to analyze the survey data, to make assertions about the population as a whole, and to take action accordingly, how wrong am I willing to be?" The answer to this question plays a key role in determining the size of the sample.

Generally, the bigger the sample size, the more precisely the data will reflect the traits of the population in question. Yet "there is no point in utilizing a sample that is larger than necessary; doing so unnecessarily increases the time and money needed" (Powell 1991). Detailed information on selecting a sample, in addition to clear explanations and useful guidelines for the sample surveying process, is provided in a number of publications listed in the Works Cited; some specifically address surveying in the library context.

Once the survey methodology has been determined, the data collection instrument, which in essence is the database forming the heart of a survey project, can be designed. The use of database or spreadsheet software, or a combination thereof, for data collection is highly preferable to paper and pencil, because it facilitates the gathering, manipulation, analysis, and portability of the survey data. Employ software that is familiar to the surveying staff and that does not encumber the data collection and analysis process. The database should be flexible in design in order to accommodate any changes incurred along the way; it should also be compatible with other working databases used by the institution for collection management. Furthermore, it is important to design the database with an eye towards the future. When the formal audio survey is complete, the database may be used to record information about incoming audio collection materials. It also may be expanded to catalog new preservation masters and listening copies. Finally, some of the survey data related to original recordings can be repurposed as technical metadata describing the source recording when original materials are digitally reformatted.

The data collection instrument should be tested before the survey is formally underway. This pretest serves as an opportunity to identify any aspect of the survey that may have been overlooked, to refine the tool, and to streamline its use. A pretest is also useful for estimating the time required to survey a given number of items and helps surveyors to hone their senses and achieve a baseline understanding of what to look for and what they can expect to find when the survey is underway.

Adequate workspace, certain tools and supplies, and established handling protocols are required for survey work. At a minimum, a sturdy table with ample surface area is required so that, for instance, phonodiscs can be safely removed from their jackets and sleeves for inspection of their condition and identification of their composition. Materials may be cracked or broken, so housings should be opened and items removed with great care (for more information on sound recordings handling, see Warren 1994). Cotton gloves should be available for handling recordings made of delicate materials and conservation instruments such as spatulas and tweezers may come in handy, as will a good source of light when close inspection of materials is called for. If surveying is done in collection storage areas, a laptop computer can be used. The use of a handheld computing device has been demonstrated to be effective for data collection (Drewes and Robb 2000), especially if the survey does not involve a great deal of data entry per collection item.

Information to Collect

While an audio survey may involve gathering basic or supplemental descriptive information, the survey should be primarily geared toward identifying conditions and traits, including conditions that pose a threat to the collection as a whole, the condition of one artifact that poses a threat to itself or to other artifacts, and conditions or traits that may have a bearing on preservation reformatting decisions. Conditions that may pose a threat to the collection as a whole include such things as: the temperature and relative humidity of the storage environment; cleanliness of the storage area; how and where the audio materials are positioned while in storage (which almost universally should be vertical); emergency response readiness; and general assessment of the materials' housing (adequate support and protection from light and dust). Possible conditions of an artifact that may pose a threat to itself or to other artifacts are: contamination, such as mold; extreme media deterioration, such as acetate tapes afflicted with vinegar syndrome; and extreme damage, such as cracking, deformation, or delamination of composite structures, such as lacquer discs. Conditions or traits that may have a bearing on selection for preservation reformatting may include: the uniqueness or rarity of the recording; the uniqueness or rarity of the format; variable properties intrinsic to format, such as tape thickness or base material; physical condition relative to similar items within collection; sound quality; and high use.

The notion of high use in the context of archival sound recordings is one that bears further discussion. What does "high use" mean in the case of an uncataloged collection of obsolete formats with no correlating playback equipment, a collection that, from a user's perspective, might as well not exist? In the research library context, the use of a sound recording is not the same as the use of a book; to make judgments about use as if the two were equivalent would be unfair, because in the past, sound recordings in most collections have been undervalued, marginalized research materials. Therefore, past and present usage statistics, if they are available, may be misleading at best. Estimations of past use—whether use by members of the institution's user community or by the original owner, creator, or custodian of the recording prior to its acquisition—can be surmised by examining the artifact for tell-tale signs of wear. This is where the surveyor's careful eye weighs in on the all-important matter of use. Heavily used grooved sound recordings are usually easy to identify by their dull finish, scratches, and pits. Other recordings, especially magnetic media, can be more difficult to read for signs of high use. In these cases, the surveyor turns her attention to the housing.

Housings speak volumes. The cheap cardboard and plastic used in the manufacture of the majority of original sound recording containers do not wear well. Plastic becomes scratched and cracked; the cardboard becomes abraded, tears, and crushes easily. The more they are handled, the more this evidence will appear. In addition to use, housings can reveal a great deal about the recording's storage history. Tidelines on a box or warped cardboard suggests involvement in a water-related incident. Fading suggests exposure to light and heat; gaps or holes suggest exposure to particulates. Housings may bear detailed descriptive information about the recording itself, not to mention critical data on the format, dimensions, and composition of the recording media. For all these reasons, original housings should never be discarded without studying and documenting them thoroughly before doing so. Yet, despite the many useful clues a housing may bear, the surveyor must be aware that it is not uncommon for sound recordings to make their way to an archive not in their original housing. For this reason, surveyors should bear in mind that information printed on the housing may not apply to the sound recording inside.

A chart outlining information typically collected in a survey of audio collection materials is provided in the Appendix. The information to be gathered is organized into several broad categories: the Survey, the Artifact, the Recording, the Condition, Restoration/Reformatting Documentation, and Related Materials. Within these categories, data elements are grouped in narrower categories. A selection of possible values for data elements and related definitions, comments, and guidance is included.

The information provided in the chart does not represent a comprehensive list of all formats, conditions, traits, and characteristics that may be encountered in the process of surveying. The data elements and correlating values simply represent those that describe sound recordings commonly found in the collections of research libraries, including both special collections and general collections. Highly-specialized collections may contain unusual formats that do not appear on this list. Collections of published sound recordings, particularly out of print material, may call for a higher degree of documentation concerning contents and rights information. The chart is intended to serve as a starting point from which survey designers can develop a set of data elements that adequately addresses the range of possibilities reflected in the collections to be assessed. Many of the data elements may have more than one value and are therefore repeatable. Though not included in the chart, the possible values of "unknown" or "other" may be included as values in the data collection instrument, because while surveying it is not uncommon to come across a collection item whose characteristics are not easily identified or assessed or does not neatly match any of the predetermined elements and values.

Taking Action Based on Survey Results

Once accurate data has been gathered about the audio collection, including the range and volume of format types and the scope of preservation problems observed, the institution is poised to take action. It is likely that critically necessary preventive preservation measures will be readily apparent, such as the need for improved housing or storage of materials. Yet the survey results can be leveraged to initiate broader preservation programmatic activities. With statistics in hand, librarians and archivists can hold meaningful discussions with their fellow stakeholders and other individuals who have an interest in preserving the institution's sound recording holdings. These stakeholders include members of the immediate user community who can provide input concerning the research value of the surveyed audio materials, therefore contributing to the process of setting priorities for preservation attention. With an understanding of the formats in the collection with the greatest needs and highest research value, consideration can be given to whether the institution should acquire correlating playback equipment. Audio specialists, such as transfer engineers with experience in working with archival materials, can play a very important role by helping to sort out appropriate reformatting solutions and their costs. Archivists and librarians at other institutions who are or have been involved in audio preservation activities can serve as another invaluable resource, as their experiences may offer a pragmatic perspective on preservation reformatting projects involving specific formats as well as the impact of reformatting on public service and the overall continued need for collection management, especially if digital reformatting is involved. Whatever pathway is followed, the likelihood is high that an audio survey project will raise general awareness about audio preservation within and throughout the institution and the community it serves, often leading to an increased level of support—tangible support, not simply moral support—from upper administration to make audio preservation a routine programmatic activity. When it is clear from the survey data which sound recordings are at stake if the risks they face remain unaddressed, then the incentive to take action in order to prevent their loss is realized.

Appendix: Survey Data Elements for Sound Recordings

	Data Elements	Possible Values	Comments
About the Survey	Date	[free text]	It is useful to maintain documentation related to the data collection process itelf.
	Name of Assessor	[free text]	
	Record Number	[automated]	
About the Artifact			
Collection Information	Collection Name	[free text]	
	Accession Number	[free text]	
	etc.		
Location Information	Range/Shelf number	[free text]	
	Box Number	[free text]	
	Item Number	[free text]	
Identification Information	Artist/Creator	[free text]	
	Title	[free text]	
	Orig. Recording Date	[free text]	Establish and use a convention for the syntax, possibly conforming to a standard.
	Item Recording Date	[free text]	A given artifact may have been created later than the date the sound it bears was originally recorded.
	Publisher	[free text]	
	Place of Publication	[free text]	
	Relation to Parts	example: "2 of 5"	Often a single sound event is represented on multiple items.
	Role of Item	original	
		master	
		edit master	
		preservation master	
		dub master	
		use copy	
		commercial	
		commercial dub	
		original dub	
	Program Duration	[hh:mm:ss]	Consider writing temporal duration of the recording according to a ISO standard syntax.
	Format	Open Reel Tape	This is by no means an exhaustive list. Recorded sound has taken many forms in its 125-year history. Among digital audio formats alone, already there are a number which have come and gone. Refer to the Audio Identification Guide by Sarah Stauderman for more detailed information about sound recording formats. Remember that videotape formats have a history of being used for holding audio information in archives.
		Compact Cassette	
		Mini Cassette	
		8-Track	
		DAT	
		Beta	
		VHS	
		U-matic	
		Optical Disc	
		Phonodisc	
		Sound card	
		Cylinder	
		Wire	
		Belt	
		Magnetic Film	
		Optical on Film	
		Magneto-Optical	
	Genre	Music	
		Field Recording	
		Spoken Word	
		Oral History	
		Natural History	

Data Elements	Possible Values	Comments
About the Recording		
Tape		
	Soundtrack	Many of these elements apply to sound on film, too.
	Broadcast	
Stock Brand	[free text] example: Scotch 111	Caution: Different brands/formulations/compositions/ages of tape may be spliced together on single reel. Caution: Tape reels and cassettes often do not make it to an archives in their original housings, so stock info on boxes or reel flanges may be inaccurate. Some brands from certain periods are known to be problematic, e.g., Ampex reels from 1980s are prone to Sticky Shed Syndrome.
Base Material	polyester (Mylar)	Mylar stretches before breaking under stress. Introduced in 1957. The most stable base material.
	cellulose acetate	Identification tip: light is transmitted through a wound reel of acetate tape. Acetate emits vinegar odor if hydrolysis is underway. Used in tapes until the mid 1960s. Should be considered a preservation priority.
	polyvinyl chloride	PVC can be difficult to identify; burns green (presence of chlorine). Used in the earliest audiotapes.
	paper	Paper tape tears very easily; it becomes brittle over time and, if exposed to water, can block up. Should be considered a preservation priority.
Gauge	1/8 in	
	1/4 in	
	1/2 in	
	1 in	
	2 in	
	3 in	
	1.25 mm	
	6.5 mm	
	8 mm	
	16 mm	
Oxide	gamma ferric oxide (Type I)	According to Van Bogart, magnetic properties of Type I and III are more stable, but Type II and IV are capable of higher output signal and higher frequencies.
	chromium dioxide (Type II)	
	cobalt-modified gamma ferric oxide ME (Type III)	
	metal particulate (MP) (Type IV)	
Thickness (mil)	0.5	While it is commonly understood that audio reel tape comes in three standard thicknesses, in fact some brands do not conform precisely to the industry conventions; thus it is possible to find tape of 1.4 mil thickness, for example. In any case, it is useful to document a tape's thickness (if not exactly, then approximately) because thinner tape is more prone to physical and sonic problems than thicker tape. It is possible to train your fingers to sense the thickness by touch.
	1	
	1.5	
Reel diameter (in)	3	Tape wound onto a hub without flanges is referred to as a pancake (see below).
	5	
	7	
	10.5	
	not applicable	
Pancake diameter (in)	[free text]	Pancakes require careful handling to prevent unwanted unraveling of the tape pack.
Reel length (ft)	600	
	1200	
	1800	
	2400	
	3600	
Cassette Length (min)	15	Often length of tape in minutes is indicated in the stock type (e.g., TDK SA90 is 90 mins.). The longer the cassette, the thinner the tape.
	30	
	45	
	60	
	90	
	100	
	120	
Track Format	full track	An instrument or solution designed to reveal magnetic tracks to the naked eye can be used to investigate the track

Data Elements	Possible Values	Comments
	half track	format of a tape recording, to see if the heads were misaligned when recording was made, or see if any recording has been made at all.
	quarter track	
	8 track	
	16 track	
	32 track	
Sound Field	mono	
	stereo	
	quadrophonic	
Noise Reduction	Dolby A	Noise reduction encoding methods should be documented because their use is essential for accurate playback.
	Dolby B	
	Dolby C	
	dbx	
Tape Speed (ips)	15/16	The slower the speed, the longer the recording's duration. In other words, sound recorded at 7 1/2 inches per second requires less physical tape media than sound recorded at 15 inches per second. Often amateur recordings were made at slower speeds in order to use tape more efficiently.
	1 7/8	
	3 3/4	
	7 1/2	
	15	
	30	
	variable	
Phonodisc		
Diameter (in)	5	
	7	
	8	
	10	
	12	
	14	
	16	
	19	
Rotation Rate (rpm)	80	See Disc Rotation Rate Note below.
	78	
	33 1/3	
	45	
	16	
Rotation Rate Note	[free text]	Rotating speed affects pitch of sound. 78s were not always recorded at true 78 rpm. Use this field to indicate known variation.
Composition	paper	Early discs may be composed of various combinations of these materials. Look in center spindle hole to determine support to which lacquer is adhered (if brown, then support is made of cardboard; silver means a metal support, while clear indicates a glass support). Acetate discs often have a bluish color along the outer rim. A disc with a cardboard core is light, while one with a glass core is heavy. Rubber includes a range of possible materials, such as vulcanite.
	rubber	
	wax	
	metal	
	shellac	
	glass	
	acetate	
	vinyl	
Side Layout	1	Some phonodiscs have sound recorded on one side only, while some titles are comprised of multiple discs.
	2	
	3	
	4	
Groove	vertical (hill & dale)	
	lateral	
	center-start	
	microgroove	
Stylus Size (mil)	[free text]	Appropriate playback of grooved recordings requires the appropriate stylus. Styli are available in range of sizes between 2 and 5 millimeters.
Stylus Type	Truncated elliptical	
	Spherical	

	Data Elements	Possible Values	Comments
		Elliptical	
Optical Disc			
	Format	CD-A	For identification tips, refer to *Care and Handling for the Preservation of CDs and DVDs -- A Guide for Librarians and Archivists* by Fred R. Byers and published by the National Institute of Standards and Technology and the Council on Library and Information Resources, 2003. Available for download at this URL: http://www.itl.nist.gov/div895/carefordisc/CDandDVDCareandHandlingGuide.pdf.
		CD-ROM	
		CD-R	
		CD-RW	
		CD-Plus	
		DVD-ROM	
		DVD-R	
		DVD+R	
		DVD-RW	
		DVD+RW	
		DVD-RAM	
		MiniDisc	
	Stock Brand	example: "Mitsui"	
	Record Process	molded	
		organic dye	
		phase-changing film	
	Metal Layer	Gold	
		Silver	
		Silver alloy	
		Aluminum	
		Silicon	
Cylinder			
	Brand / Composition	Brown wax	Brown wax is susceptible to mold and oils from human skin; both brown and black wax are susceptible to deformation caused by temperature changes.
		Black wax	
		Blue Amberol	
		Purple Amberol	
		Pink Amberol	
		Black Amberol	
	Core Composition	plaster	Later cylinders have a supportive core of a different material.
		metal	
		paper	
		none	
	Interior Diameter (in)	[free text]	
	Cylinder Length (in)	[free text]	
	Rotation Rate (rpm)	90	
		120	
		160	
About the Condition			
Visual Inspection of the Artifact			
	Contamination	none	
		dust	
		fungus	Mold can grow on tapes, early cylinders, and paper housings, such as phonograph sleeves.
		water	
		chemical	
	Odor	none	
		musty	
		vinegar	Vinegar odor indicates hydrolysis of cellulose acetate. Other distinctive odors may indicate binder hydrolysis.
		dirty socks	
		waxy	
	Housing	cracked	A housing can reveal much about a recording's use and storage history, but beware: not all housings are original to the item they contain, so they cannot always be trusted.
		stained	

Data Elements	Possible Values	Comments
	gaps	
	high use	
	water damage	
	torn/ripped	
	no damage	
	no housing	
	not original housing	
Cassette/Reel	cracked	
	loose parts	
	detached pressure pad	A cassette may have a detached pressure pad, rendering it extremely vulnerable to damage if played back.
	broken hub	
	detached tape from hub	
Tape	step pack	With the possible exception of collections of professionally-produced tapes, a tape with a smooth, even wind is a rare thing in archives. Therefore, noting the condition of the wind may not be useful for prioritizing items for reformatting, yet it is worthwhile to record the condition as it indicates a sense for the overall condition of the collection. For definitions of these conditions typically exhibited by tapes in archives, see Van Bogart.
	flange pack	
	popped strands	
	blocking	
	cinching	
	cupping	Cupping describes tape that curls such that the edges are not in plane with the tape surface.
	edge damage	Edge damage can occur when tape strands have "popped" from the wind and rub against other objects, such as the reel flange.
	spoking	Spoking occurs when the tape wind is such that lines appear radiating from the center hub like the spokes of a wheel, and the tape itself begins to take the shape of a polygon rather than a circle.
	splices	Splices may introduce a number of potential problems in the stability and playback of a tape.
	oxide shed	The oxide of cheap/early magnetic tape can delaminate from its base, sticking to the adjacent wrap. Also, tape with binder hydrolysis may exhibit sticky shed syndrome when played back, a condition where the oxide sheds onto the read heads, leaving a sticky mess and effectively destroying the recorded sound.
	stretch	Tapes can become deformed if they are stored for an extended period with a poor wind.
	object in wind	It is not uncommon to find an object, e.g., a scrap of paper, wound into the pack to mark a place in the recording, like a bookmark.
	loose wind	
	tight wind	
	white powder	May indicate binder hydrolysis.
Discs and Cylinders	ooze	May indicate deterioration of splice adhesive or compound in tape binder.
	scratched	Many of these values apply to traditional phonodiscs, cylinders, and optical discs alike.
	cracked	
	broken	Many discs come to archives broken. Shellac can shrink slightly over time so pieces may not fit together neatly.
	chipped	
	pitted	
	powdery substance on surface	Plasticizers in the lacquer coating of discs can leech over time, forming a whitish powder on the disc surface.
	dull from wear	A grooved sound recording that has been played often will have a dull finish, if the scratches do not already give it away.
	flaking / crazing	Acetate lacquer will craze as it becomes brittle and shrinks, ultimately delaminating from its support.
	warped / misshapen	
	bronzing	Bronzing, or "bit rot", is oxidation of the reflective metal layer.
	label delamination	
	hazing	Hazing is a lack of clarity in the top (polycarbonate) layer, often occurs from the outer edge inward and may be caused by act of writing on surface or penetration of solvents in pen ink used to label optical media.
Wire	tangles	
	breaks	
	rust	Early wires not made of stainless steel may rust.
	snarls	

Aural Inspection of the Recording

Data Elements	Possible Values	Comments
Sound characteristics	wow / flutter	"Waver in a reproduced tone or group of tones that is caused by irregularities in turntable or tape drive speed during recording, duplication, or reproduction" (Encyclopedia Britannica 2003).
	print through	"The condition where low frequency signals on one tape winding imprint themselves on the immediately adjacent tape windings" (Van Bogart 1995).

Data Elements	Possible Values	Comments
	head misalignment	Tape equipment with misaligned heads will not record the signal optimally, failing to make full use of recordable media and track layout. Misalignment is a common problem in cassettes.
	distortion	Distortion, a broad term used to describe the effect on a sound wave which has been mechanically or electronically altered in an unintended or unanticipated way, can result from a number of causes, such as a stylus ill-fitted to a disc groove during playback, a poorly placed microphone in a performance, or aliasing from improper digitization.
	low signal strength	A recording produced at low levels is subject to distortion in the re-recording process.
	hum	The sound system itself can introduce an audible hum or background noice to a recording made on that system.
	tape hiss	Constant, even noise underlying the signal resulting from "the inability of the recording system to organize completely the magnetic [particles of audiotape in a recording]" (Encyclopedia Britannica 2003).
	squeal	High-pitched noise exhibited by tape suffering from lubrication loss or Sticky Shed Syndrome.
	surface noise	Unwanted sound introduced by the action of the stylus physically moving across a phonodisc surface.
	rumble	A type of distortion, rumble is "the sound produced by vibrations in the recording or playback turntable of a disc system" (Encyclopedia of Recorded Sound in the United States 1993).
	drop outs	Signal loss caused by missing magnetic material or obstacle preventing proper tape-to-head contact.
	groove damage	Scratches, pits, and other imperfections in the disc surface and groove shape may result in noise such as pops and
Playback Note	[free text]	Any additional information pertaining to the playback of the item should be documented.
Restoration / Reformatting Documentation	[free text]	Any information pertaining to the reformatted or restored copies of the original should be documented.
Related Materials	[free text]	It is not uncommon to find paper documents and other items stored in audio recording housings. They should be stored separately from the sound recording itself but it is important to document any relevant information they contain and to maintain their intellectual relationship with the audio material.

Works Cited

Byers, Fred R. *Care and Handling for the Preservation of CDs and DVDs: A Guide for Librarians and Archivists* . Gaithersburg, MD: National Institute of Standards and Technology and the Council on Library and Information Resources, May 2003. <http://www.itl.nist.gov/div895/carefordisc/CDandDVDCareandHandlingGuide.pdf> (29 September 2003).

Encyclopaedia Britannica Online. 30 Sep 2003. <http://search.eb.com/eb/> (30 September 2003).

Marco, Guy A., ed. Encyclopedia of Recorded Sound in the United States. New York, Garland Pub., 1993.

Van Bogart, John W.C. *Magnetic Tape Storage and Handling: A Guide for Libraries and Archives.* Washington and St. Paul: The Commission on Preservation and Access and National Media Laboratory, 1995.

Acknowledgements

The author gratefully recognizes the support and contributions of Cathy Aster, Connie Brooks, Maria Grandinette, Walter Henry, Richard Koprowski, David Seubert, and Sarah Stauderman.

Works Cited and Other Sources

Allen, David Randel. *State of Recorded Sound: Survey of Surveys*. Report prepared by the Communications Office, Inc. for the Council on Library and Information Resources, 2003.

Babbie, Earl R. *Survey Research Methods*. 2d ed. Belmont, CA: Wadsworth, 1990.

Busha, Charles H. and Stephen P. Harter. *Research Methods in Librarianship: Techniques and Interpretation*. New York: Academic Press, 1980.

Creswell, John W. *Research Design: Qualitative and Quantitative Approaches*. Thousand Oaks, CA: Sage, 1994.

Drewes, Jeanne and Andrew Robb. "The Use of Handheld Computers in Preservation and Conservation Settings." Paper presented at the annual meeting of the American Institute for the Conservation of Artistic and Historic Works, Philadelphia, June 2000. Information available at http://www.lib.msu.edu/drewes/Presentation/palmp/survey.html and http://www.lib.msu.edu/drewes/Presentation/palmp/handout0606.doc.

Drott, M. Carl. "Random Sampling: A Tool for Library Research." *College and Research Library News* 30 (1969): 99–125.

Powell, Ronald R. *Basic Research Methods for Librarians*. 2d ed. Norwood, NJ: Ablex, 1991.

Seubert, David. "Designing and Managing an Audio Preservation Program." Paper presented at the annual meeting of the Association for Recorded Sound Collections, Philadelphia, May 2003.

Smith, Abby. "Background for 14 March meeting." E-mail to Connie Brooks. 1 March 2003.

Warren Jr., Richard. "Handling of Sound Recordings." *ARSC Journal* 25 (Fall 1994): 139–62.

Risk Reduction through Preventive Care, Handling, and Storage

Alan F. Lewis
Subject Area Expert—Audiovisual Preservation
Special Media Archives Services Division
National Archives and Records Administration

WHEN I THINK ABOUT RISK to a collection of machine-based AV archival materials, I think about:

- loss, damage, or destruction of the physical item, and also
- loss of access to the content of the recording, the information content.

Hence, our risk reduction thinking needs to have a wider scope than just the stuff on the shelf in the back room. As I view risk reduction, then, I try to think about the recording media as parts of an AV recording system, because without the system, the item on the shelf is of little use. Reducing risk, then, means reducing the risk to all the components of the system.

My assigned topic is to discuss care, handling, and storage:

▶ To me, *care and handling* means understanding the various recording media and their matching equipment in a collection and using appropriate techniques to eliminate or minimize potential damage to those media and to the equipment when the former, the recorded media, is off-the-shelf and being used and the latter, the equipment, is turned on and in operation.

▶ Likewise to me, *storage* means understanding the various recording media and their matching equipment in a collection and using appropriate techniques to eliminate or minimize potential damage to those media and to the equipment when the former, the recording media, are in containers and on the shelf and the latter, the equipment, is turned off and dormant.

So to understand these things, permit me to do some basic training first and then briefly review my "19 Conservation Concerns."

Machine-based audiovisual recording systems are composed of three elements. The first and most obvious, because it's the "stuff" on the shelf, is the recording media. It is the physical item that has been used to *fix* in some [hopefully] permanent way the sounds that are the collection, whether those sounds are music, spoken words, or natural sounds. For audio recordings, the media have had a variety of shapes: cylinders, flat discs of various thicknesses, endless belts, wires on spools, and ribbons of material running from reel-to-reel in the open or enclosed in housings.

The second element of a recording system is the equipment, the devices that initially captured and later can retrieve the sounds that have been fixed on the recording medium. Over the century plus that sound recordings have been in existence, the equipment has used *transponders* to convert sound waves into either mechanical energy or electrical energy. That energy, in turn, drove styli connected to diaphragms or electromagnets. Edison's original acoustical cylinder machines, for example, used a horn-diaphragm-stylus-soft cylinder as the technology.

The third element is the standards that were developed as a part of the invention of the system. They specify all the details of how the signal passes through the technology, onto the recording medium and later, how they are retrieved in usable form. Again, using the example of a cylinder machine, the standards involved the dimensions of the cylinder, its speed of rotation, the relative softness of the surface to be incised, the number of grooves per inch that the device would cut across the surface of the cylinder, etc., etc.

Now, because I think it is very important to understand the recording media, let me get into a bit more detail.

A typical audio recording medium is likely to have at least two of the following components:

- a base or substrate
- a signal capturer
- a linker

It is important to understand that each of these components may have *natural enemies* which would include their own built-in seeds of deterioration, sometimes called *inherent vice* in the archives field, as well as *unnatural enemies,* the conditions we subject the media to over time: its use, overuse, and misuse.

The first component is the *base* or *substrate*. It is the physical foundation of the recording medium. It typically has physical size (dimensions) and a shape. It is a cylinder, a disc, a wire, or a tape and generally the term we use to describe the medium is the physical form of that substrate. ("How many discs do you have in your collection?")

The second component is the *information capturer*, the technical system by which the information is transformed from sound energy in the air into whatever means is used by the system to fix it on the medium: variable grooves for cylinders and disc recordings and electromagnetic impulses for the magnetic media.

The third component, lacking in some recording media, is a *linker*, the physical or chemical means by which the information capturer is secured to the substrate. In original phonograph records, one-off instantaneous *discs*, the linker is the bond between the cellulose lacquer layer into which the grooves are cut and the base material which might be cardboard, aluminum, glass, etc. With legacy magnetic tape materials, the linker is typically a urethane plastic binder. (Modern mass-produced phonograph records have no linker because the grooves are impressed right into the substrate. Likewise, magnetic wire has no linker because the wire itself, the substrate, becomes magnetized.)

Bear in mind that because sound recordings come in so many types, I can't go into any real detail about any of them in the short period of time I have. However, they include:

- acoustically made mechanical recordings
- electrically made mechanical recordings
- magnetic wire recordings
- magnetic recordings on paper, cellulose, and polyester base tape

With all of these, knowledge may get more difficult because there are original, *one-off recordings* and also *mass-manufactured products*. The bottom line is that if humans created these technologies, the standards and the recordings, something can and will go wrong with them over time. What can go wrong are what I call the "Attacks Against Recording Systems."

First, with the recording media, there is their own *inherent vice*, the deterioration factors that are basically manufactured into the media because they weren't manufactured for the long term. One might think of it as *natural deterioration* or *natural aging* because most media were designed for short-term, commercial use with little or no thought for their long-term keeping qualities. Secondly, because these media are handled by people and by machines, there is the wear and tear of overuse and mishandling by the uninitiated, the careless, or the mean-spirited.

Attacks against the technology begin with the real life fact of commercial obsolescence of technologies. Technologies give way as newer systems come on the market that are more desirable and therefore more acceptable. (Consider, for example, how computer removable storage technology has gone from 8" floppy discs to 5¼" floppies to 3½" floppies to CDs.) Consider, too, that these technologies we use were more often designed for production and/or distribution purposes without thought to their long-term availability. Finally, there are the costs and ultimately the impracticality of equipment upkeep as technicians and spare parts become unavailable and as skilled operators retire or die.

Standards (and software if we're in a digital domain) are also subject to obsolescence as new versions are developed and the old ones are consigned to the audiovisual scrap heap. In cases where new versions are marketed, the question of backward compatibility arises as version replaces version. Consider with grooved discs alone, how standards have changed with cutting inside-out to outside-in tracks, rotation speeds from 16-2/3 rpm and 78 rpm to 45 and 33-1/3 rpm (to say nothing of half-speed mastering!), and groove pitch from old standard (100 grooves/inch) to microgroove's (200 grooves/inch).

Finally, there is the legitimate concern about quality loss through signal compression that discards information (lossy compression) and even what can happen with compression that doesn't dispose of material (lossless compression) but may leave audible artifacts as a result of the electronic processes.

In summary, in my view, risk reductions starts by knowing the details about the media, the technology, the standards, and the software, and paying attention to those details.

Now, let me move on to my "19 Conservation Concerns" because whether handling the media or storing it, these are relevant. Again, because of the complexity of the audio recording field, much of this needs to be general and not medium- or system-specific.

Environment. Conservators in all fields, not just AV media, cite the temperature and humidity conditions in which heritage media are stored as the single most critical factor in their long-term survivability. If materials held by an archive are not to be at-risk, proper levels of temperature and humidity are required. Because of the diversity of sound recordings to be discussed in this conference, I cannot make any single recommendation but rather suggest that each archive consider the variety of media it holds, do the research on a medium-by-medium basis, and design storage environments that are appropriate for each. I will say that at present, the U.S. National Archives holds its wire, acetate, and polyester magnetic media and its mechanical media at 65 degrees F @ 30% RH. The environmental control system must operate "24/7/365" and must be monitored and recorded continuously in order to have proof-of-performance of the system.

Physical Security. It is a given that archives have the responsibility to provide various kinds of protection for their accessioned materials. Risk reduction, therefore, requires protection from theft, vandalism, and damage as well as protection from unauthorized duplication and use. In a legal sense, risk reduction might also be expanded to include making sure to observe and honor the copyright of recordings, enforcing donor and privacy restrictions, etc.

Fire Protection. Allied to physical protection is eliminating the risk of fire through proper selection of the archive site as well as having a fire prevention program in place, an effective fire detection and warning system, a fire suppression system that minimizes collateral damage, and having a recovery plan in place to minimize loss to media, equipment, and administrative records after an emergency.

Water Protection. Also allied to physical protection is reducing the risk of water emergencies by site selection to defend against acts of nature, eliminating all overhead piping other than possibly a water fire suppression system, and maintaining a sound building envelope, especially roofs, windows, and skylights. A water detection system should be installed and a water emergency recovery plan should be in place and include provisions for dealing not only with the media but also the equipment, the finding aids, and the administrative records. Some additional practical hints include not using basements or attics for storage, building above the calculated 100-year flood level, training the staff not to store media on the floor, using storage furniture with the bottom shelf at least four inches above the floor level, and providing floor drains.

Light Sensitivity. Other than new media discs that use light sensitive dyes, I know of no audio media that are particularly light sensitive. On the other hand, ultraviolet radiation from sunlight and artificial lighting sources are known to be problematic to paper labels, adhesives, and inks. Reduced level lighting also has a positive effect on electric utility costs, both from the cost of the illumination itself as well as the cost of removing the heat caused by electric light sources that increase the heat load the environmental system has to remove.

Cleanliness. Another logical risk reduction step is to store and use sound recordings and their equipment in clean locales for two very good reasons: the presence of foreign matter on the playing surface of any audiovisual medium produces a reduced quality playback, and the introduction of dirt and other physical debris into the equipment also gets on the media that pass through it. Hence, common sense good housekeeping measures should be in effect in all collection, storage, handling, and use areas.

The traditional "no smoking/eating/drinking" rules always apply. Unpacking collections should take place away from storage and use areas so their "street dirt" isn't introduced into those areas. Hard surfaced flooring rather than carpeting is desirable to make cleaning easier. Surface dusting and floor cleaning should take place on a regular basis and should be conducted without solvents or any other substances that have not been tested and found archivally acceptable.

Air Quality/Pollutants. Related to cleanliness is the matter of keeping gaseous pollutants out of storage and use areas. Since many audio archives are located in urban areas, their natural air supply is likely to contain sulfur and nitrogen compounds and ozone. Man-made materials used in building components, in insulation, and in furniture can introduce formaldehyde. Cellulose acetate products bring the product of their own deterioration, acetic acid, into the space. Risk reduction factors should therefore include selection of the site for the facility, careful selection of materials to be used in it, sophisticated air filtration systems, and measuring and tracking pollutants on a regular basis.

Biological Infestation. We often observe mildew problems with sound recording collections, especially those that have been stored in less than optimum environments. We usually don't think of such collections as being at risk from insect or rodent infestation. In reality, though, our audio recording media and/or their packaging are subject not only to the "no-see-um" spores that bring mold but also the "macro-critters" like insects and rodents that thrive on the cardboard packaging, paper labels, and adhesives.

Risk reduction techniques include using proper temperature and humidity levels for the preservation of the materials which, coincidentally, are levels that do not provide a hospitable environment for the critters. Pre-storage inspection and fumigating collections may be needed—but careful and knowledgeable fumigation so that the process to rid the collection of the problem doesn't damage or destroy the collection itself! Good housekeeping, of course, will remove the other enticements of food or water that attract the pests. Finally, if there is a problem, use an integrated pest management system that first identifies the specific problem and seeks to solve only it. This avoids introducing broad-spectrum pesticides that may be more than are needed.

Strategic Dispersal. Strategic dispersal sounds like a Department of Defense term, but in audio risk reduction terms, it simply means that if there are multiple copies of an item, do not store them together. The point is, if there is a destructive emergency situation, hopefully one or the other copy will survive. Dispersal might mean putting copies on opposite sides of the same room, in different rooms, in different buildings, etc.

Primary Containers. In practice, there are as many different types of containers on the market as there are recording media and there are variations based on manufacturing cost. All too often, the primary container that accompanies the purchase of blank stock or a new pre-recorded item, is not designed for the long-term keeping of the item. Again, because of the many audio recording media we deal with, time does not permit me to go into detail about any one container. There are considerations to be given to design, construction and materials used, the chemistry and stability of that material so it is non-reactive with the medium, and whether it is equipped with media support devices. I prefer containers for all media that have a positive closing device, an adequate writing/labeling surface, and a design that resists water penetration. A one-piece item with its body and cover permanently attached is best. If the container is plastic and the collection is a large one, some local fire codes may require a flame retardant be used in the material to lessen the danger to firefighters. Toxic gases are released from some plastics when they burn. Finally, some media, especially acetate materials, may do better in a ventilated container than in a tightly closed one.

Storage Position. Proper positioning for audio recording media has long been recognized as an important risk reduction activity. The risk is generally related to the force of gravity pulling downward on media over long periods of time. For the linear media, wire recordings seem to pose no problem whatever their orientation is in relation to the force of gravity. Tape, on the other hand, should be stored vertically in order to protect the edges of the medium from deformity and edge damage problems caused by pressure against the reel's flanges.

Discs are traditionally stored vertically to prevent warping and to prevent stacking too much weight in a pile of discs and causing breakage in the lower ones. However, in my opinion and based on observing what I believe to be stress fractures in some shellac-type pressings, the archivist should consider the various types of discs in a collection and make reasoned positioning decisions based on understanding the nature and structure of individual types of items. Glass-base instantaneous discs are very different from vinyl pressings and some should and could be stored vertically and some horizontally.

Winding Ribbon-Like Media. The winding of recording media is relevant only to the linear media like magnetic wire, open reel tape, and tape in cassettes and cartridges. With wire, there is a risk that during play or shuttling from spool-to-spool, it will slip over a flange or backlash resulting in what one laboratory person describes as ending up looking like a bird's nest. For the tape media, the results of poor winding are possible snapping of acetate tape and stretching of polyester tape.

One school of thought on tape winding for storage suggests playing the tape through from end-to-end and leaving it unrewound. Assuming the machine is properly adjusted and aligned, this should result in a smooth and uniform tape pack with even tension throughout. Leaving the tape in this *tails out* orientation means that should there be *print through* of the signal, that problem will be audible as an echo with the ghost signal following the strong signal. If left *heads out* and print through develops, it will be a *precho* (this is a made-up word, a contraction for "pre-echo") with the ghost signal heard before the strong signal which is more distracting to the human ear and human mind.

A frequently asked question has to do with the need for periodic winding of linear media like wire and tape. There seems to be as many schools of thought as there are audio archivists and their opinions include:

- Do periodic rewinding every "x" number of years because tape is designed as a flexible medium and it should be "exercised." This approach might mean instituting a program to rewind the entire collection over, say, five years and therefore doing 20% of the collection each year so that in the five-year cycle, everything is rewound. A subset of this would be to do a visual inspection first of that 20% to see if there is any evidence that winding is necessary.
- Another school of thought suggests, "If it ain't broken, don't fix it," meaning leave it alone. This thinking may be based on the fact that any running of tape on a machine potentially puts the tape at risk from handling, wear and tear, and machine malfunctions.
- Yet another thought is to leave the tape alone unless there has been some major environmental change in the storage area that might have caused the tape to expand or contract due to temperature and humidity changes which resulted in changes in the tension of the tape packs.

Shelving. Another aspect related to physical protection has to do with the shelving used for the storage of media. Risk reduction associated with shelving include using archivally acceptable materials, shelving that is robust and proper for the size and weight of stored media, adjustability so shelf heights can be varied as needed, shelving that does not encourage climbing, and shelving that does not allow "hiding places" behind the upright structural members. Local or institutional building codes must be observed or even exceeded and even if not required, earthquake and anti-tip devices should be installed.

In new facilities or facilities to be renovated, consideration should be given to compact shelving in order to maximize the amount of material that can be stored per square foot in the expensive to construct and expensive to operate environmentally controlled and protected vault space. Don't forget about floor loading, because compact shelving significantly increases it!

Shock/Vibration Protection. Another obvious subset of physical protection of media, and therefore eliminating risk, is protection of the materials from shock and vibration. This includes a generous helping of staff training so they do not consider these sometimes physically heavy valuable historical records, as "stuff" and handle it as such. Potential damage includes causing breakage of the recording media itself, problems with reels and cassette shells, broken containers, and the concern that major shocks might cause disarrangement to magnetic particles.

Magnetic Protection. A fortunate characteristic, or unfortunate depending on one's viewpoint, of magnetic media is that information recorded on it is erasable by the introduction of new magnetic information or by subjecting it to a stronger magnetic field. This was the marketing virtue of magnetic audio media: it could be erased and reused; but it now becomes the curse of magnetic media historians and archivists, the loss of potentially valuable recordings in order to reuse tape stock. Hence, the risk is purposeful or accidental, partial or complete, erasure. Risk elimination includes staff and user training, disabling the record function on equipment if practical, disabling those media that have an anti-recording interlock, and protecting the media from strong electromagnetic fields.

Item Identification. The risk here is item *misidentification* and the potential loss of the item physically in the storage area or intellectually in the description and cataloging scheme. A rule of thumb in dealing with the situation of a recording in a container that has another number on it is to look for the container that matches the number or name of the "miscontainered" item. In most cases, what has happened is that the two recording media and their cases were inadvertently switched by some previous archivist, technician, or user.

With regard to intellectual content, there are established rules and procedures for describing audio recordings that should be used. It is possible that in some kinds of collections, especially those containing unedited or untitled items, a name or number will have to be provided by the cataloger using a system that matches the system already in use with the collection or the institution. Whatever naming system is used, it is clear that individual media and individual containers should both carry that identification, and the marking system used should be archivally acceptable.

Inventory Control. Small collections, especially those with little access activity, require little in the way of a fancy inventory control system. As collections grow, however, the problems and the risks associated with loss or misplacement of items becomes a factor that needs to be considered. Traditional archival procedures often involve assigning records series designators and numerical sequences within series and such procedures may work with some types of audio collections. Be warned, however, that complexity continues to grow as the archive accessions multiple production generations of the same item and related media that may require different storage environments. Multiple copies, accessioning related materials out of numerical or chronological order or over a span of time, different sized storage containers, and possibly different storage positioning requirements, all add to the mix.

Consequently, in my opinion, the answer is to take a page from modern warehouse practice and develop a random location storage system. Such a system is one that takes into consideration the variable factors of environmental needs, size, positioning, etc. and assigns items to the next available shelf space that meets the requirements. Computer software can be developed or may already exist that can take into account all those variables. Bar codes or other automatic identification systems (AIS) can be used to identify the items as well as identify the shelf locations so the entire process can be virtually error free because it does not depend on frequent keyboarding of information.

Equipment. It is obvious that equipment is necessary to deal with machine-based AV media. Without equipment the archival functions of appraisal, media inspection, storage preparations, description, reference, user service, content preservation (reformatting), and customer reproduction needs cannot be served. As risk reduction measures, there are technical requirements for the equipment: it must be technology that matches the media formats, it should be high quality professional equipment, operators for the more sophisticated technologies must be available, and the equipment must be maintainable. The options are for the archive to become an operating machine museum or finding and contracting with an operating machine museum, usually a commercial, professional duplication house.

Personnel. As with any and all human endeavors, the success of the activity depends on the selection, training, and dedication of staff members and the leadership they are given. As a function of risk reduction, they need to be trained in the legacy media they are custodians of, trained in the new media of the users, and trained and knowledgeable in the collection they are responsible for. In addition, and not necessarily associated with risk reduction, other keys to an effective audio archives staff are to insure that they are trained in customer service, something much in vogue in management circles these days, and "collaterally trained," knowing about similar collections elsewhere in order to provide a higher level of customer service.

In summary, then, audio collections are like some people's lives: full of risk. It is our job as collections managers to know the risks to our media, our equipment, our standards, and our software, and to take steps to reduce or avoid them. Risk-reduction may be a full time job in your shop!

This paper represents work carried out for a federal government agency and is not protected by copyright.

The Save Our Sounds Project

Michael Taft
Head, Archive of Folk Culture
American Folklife Center
Library of Congress

THE USE OF SOUND RECORDING EQUIPMENT in ethnographic fieldwork has been part of folklife research for well over a hundred years. In fact, the Library of Congress holds Jesse Walter Fewkes's wax cylinders of members of the Passamaquoddy tribe that he recorded in 1890—probably the first ethnographic field recordings (Fewkes 1890; Gray and Lee, eds. 1985: 221–32). Since then, ethnographers have preserved the voices and performances of countless people from virtually all the world's cultures and the American Folklife Center's Archive of Folk Culture is the largest repository in the country of this type of material. What makes the library's collection particularly valuable is that many of the field recordings are accompanied by other documents, making each recording a context-rich package of information. These documents range from notes written on cylinder housings, disc sleeves, and tape boxes, to recording logs and long narrative field notes, to correspondence and other manuscript material related to the ethnographer's fieldwork. As well, photographs, drawings, and other graphic material often accompany these documents, making the sound recordings part of a truly multi-media package of information.

There are, of course, many problems associated with the archival maintenance of multi-media collections, but among the most serious of these is the potential deterioration of all or part of these packages of information. No sound recording was meant to last forever and each recording format presents its own set of preservation issues. Likewise, paper and photographs need their own preservation treatments if they are to remain accessible to researchers. In answering these problems, the American Folklife Center has embarked upon a pilot digitization-preservation project called the Save Our Sounds Project.

Save Our Sounds is a joint initiative between the American Folklife Center of the Library of Congress and the Center for Folklife and Cultural Heritage of the Smithsonian Institution. The project is financed through the *Save America's Treasures Program* of the White House Millennium Council and the National Trust for Historic Preservation, and administered by the National Park Service (see National Park Service and National Trust for Historic Preservation n.d.). Because of legalities involved in government agencies applying for funds from other government agencies, the Smithsonian is the receiver of monies from the National Park Service and these funds are then dispersed to the Library of Congress through an inter-agency cooperative agreement. The grant totals $750,000, of which the Library of Congress share is $285,000.

This particular arrangement between three government agencies—the National Park Service, the Smithsonian, and the Library of Congress—was to a great extent uncharted territory. Establishing the details of the agreement, including procedures for fund sharing and expense reporting, was a time-consuming affair. Adding to the complicated nature of this agreement, both the Smithsonian and the Library of Congress were expected to raise matching funds in order to access the earmarked funds from the National Park Service. Here as well, agreements had to be established on procedures for shared fundraising, as opposed to the separate fundraising ventures of each institution.

The legal and bureaucratic components of this project extended from the announcement of the award in July 2000 to April of 2001, which may seem like an inordinate amount of time, but this type of inter-agency agreement was as much a pilot for possible future agreements as the project itself was a pilot for digitization procedures. Both the Smithsonian and the Library of Congress learned much about what was, and what was not, possible in a joint agreement and the Save Our Sounds Project will undoubtedly smooth the way for future shared initiatives.

The project itself called for the digital preservation of 8,000 recordings, with the Smithsonian digitizing 5,000 of these and the Library of Congress responsible for the other 3,000 recordings. What constituted a "recording" was kept fairly loose—it could be a two-minute cylinder or a two-hour tape—but it was generally agreed that the operative unit would be a single sound storage artifact. This artifact could be in any format. The Smithsonian's selected recordings were instantaneous discs and audio tape recordings. The Library of Congress cast a wider net, including wax cylinders, instantaneous discs, wire recordings, audio tape (both open reel and cassette), DAT tape, and video tape (both open reel and cassette).

The recordings selected at the Library of Congress were all from collections held by the Archive of Folk Culture of the American Folklife Center. The strategy for selection involved a number of criteria, but the guiding principle was that the American Folklife Center would save collections, rather than individual recordings. The reason for this principle is that digitizing should be seen as a part of the processing schedule of any collection, rather than as a hunt-and-pick activity. After a collection has been donated, accessioned, inventoried, stabilized and rehoused, described and catalogued, the next processing step would, in ideal circumstances, be the digital preservation of the collection.

Obviously, each of these processing steps involves decision making, and each collection demands its own customized processing plan. Regarding digitization, the first question to ask is whether the entire collection should be digitized, or only a selected portion. Given a particular collection, certain administrative files, duplicate material, or published material may not be part of the digitized collection—in essence, for each collection a decision was made as to what part of the collection was "it" and what was not "it" from the digital point of view.

I will further describe this decision making below, but the first task of the archive staff was to select the collections that would be part of the Save Our Sounds Project. As I stated earlier, there were a number of criteria for selection, ranging from the technical to the political.

Content. The *Save America's Treasures Program* required that the material to be digitized should have American content. In effect, the recordings should be of American traditions or should reflect an American perspective on folklife. This criteria was political, in that the Americanness of any part of the archive is not normally a criteria for preservation. We are the Archive of Folk Culture, not the Archive of *American* Folk Culture, and our collections policy extends to traditional material from any of the world's cultures.

Historical or Cultural Significance. All of our collections are culturally or historically significant, and it is a mug's game to distinguish between more and less significant collections. However, archive staff were aware that certain collections were in high demand, or were well known within the scholarly community, or were likely to gather a readership, once they were made accessible in digital form. This criterion was, of course, subjective and demanded that, in order to arrive at their decisions, archival staff needed to apply their experience and knowledge of the center's archival holdings, as well as the habits of researchers.

Present State of Accessibility. Another criterion involved how accessible a collection was, in practical terms, for use by researchers. As is standard practice, we do not serve original sound recordings, but many of the listening copies in our archive are of poor quality and are deteriorating. In many cases, we have no listening copies for collections that have been in the archive for years—making these recordings almost entirely inaccessible. As well, although we do serve original manuscripts and photographs, certain items may be held back from researchers because of their fragility or deteriorated condition.

Fragility and Deterioration. A major criterion was the physical state of the recordings in a collection. Some sound recordings are, by their nature, more unstable than others—and this is not always a matter of age. A 100-year old cylinder may be more stable than a 1970s audio tape.

Variety of Sound Recording Formats. Because Save Our Sounds is a pilot project, we were intent on trying out our procedures on a variety of formats. Thus, we selected collections that called for the digitization of cylinders, discs, wires, audio and video tapes, and born digital recordings.

Complexity of Collections. Again, because the project was a test of our capabilities, we chose collections of varied complexity. Some were composed of sound recordings and little else, while others were multi-media in the extreme, consisting of several kinds of recording formats, manuscripts, and images.

Diversity of Material. The project depended upon raising matching funds from outside sources. To maximize our chances of attracting donors, we understood the necessity of including a variety of kinds of collections that would appeal to different kinds of donors. It was important that we keep in mind the ethnic and national traditions represented by the collections, their genre, their region of the country, and gender issues, among other aspects of diversity.

Other Political Considerations. In the case of one collection—the Pearl Harbor Collection—which I will describe more thoroughly below, the decision was of a political and practical nature. After the September 11th tragedy, the center staff gathered to decide how we might respond. The result was a collection of audio and video recorded first reactions of Americans from around the country; these interviews were carried out by a number of ethnographers, students, and interested citizens. The project was inspired by a similar project carried out by the Library of Congress sixty years earlier. The day after the 1941 Pearl Harbor attack, Alan Lomax of the Library of Congress called on folklorists across the country to conduct man-in-the-street interviews in order to document first impressions of the event. The timeliness of these Pearl Harbor recordings—in light of the September 11th tragedy—combined with the knowledge that the Franklin and Eleanor Roosevelt Institute had an interest in the Pearl Harbor material and the American Folklife Center had recently initiated its high-profile Veterans History Project, made the Pearl Harbor Collection a good choice for the Save Our Sounds Project. Beyond its obvious historical significance, the collection would make the project that much more of interest to donors and to the media—and would give the entire project a heightened political profile that would benefit future fund-raising efforts.

No one collection received top marks in all of these criteria and deciding on which collections to choose for the project was a matter of balance and compromise—and the collective wisdom of the reference and processing staff of the American Folklife Center. With these criteria in mind, however, eight collections were chosen for the project:

Eloise Hubbard Linscott Collection. Linscott was a collector of traditional music and song in New England. She began her collecting with a cylinder recorder, graduated to a disc-cutting machine, and eventually used a tape recorder over her collecting career which extended from the 1930s to the 1960s (Baker 1979). The variety of discs that she used included aluminum discs and lacquer discs with aluminum, glass, and paper bases that varied in dimensions and quality. In addition to approximately 450 sound recordings, her collection includes over 100 photographs and 6,000 pages of manuscript. All of this material is slated for digitization. Her collection also includes a copy of her book, *Folk Songs of Old New England* (Linscott 1939), that she modified with inserted photographs and notes—all of which makes it an "association copy" of the book—and this too is part of the digitization project. The collection also consists of hundreds of pamphlets, booklets, and other printed and published ephemera that fall outside of the scope for digitization. They are not "it" as far as the Save Our Sounds Project is concerned.

James Madison Carpenter Collection. Carpenter was an American folklorist who went to the United Kingdom in the late 1920s to record sea chanteys, ballads, songs, dance tunes, and traditional dramas. He also recorded songs and narratives in the southern United States. His collection includes approximately 180 cylinders, 200 instantaneous lacquer discs, over 400 photographs in several formats, and over 13,000 pages of manuscript. All of this material, with the exception of some of the discs (which are partly transfers from the cylinders) is being digitized. Nine thousand pages of student papers that Carpenter kept from the classes he taught, however, are not scheduled for digitization. This collection has the potential of attracting researchers who have become aware of Carpenter's work (Bishop 1999), but who have been frustrated by the poor quality and general inaccessibility of the current analog copies of Carpenter's original materials. A team of British researchers has recently created an online catalog to the Carpenter collection that will greatly add to the value of the digital presentation of this material (Bishop et al. 2003).

American Dialect Society Collection. In the early 1930s, the American Dialect Society conducted recorded interviews with New Englanders in order to gather samples of dialect (Kurath 1939–43; Dialect Collection for Folk Archive 1985). The result was approximately 880 aluminum instantaneous discs and 1,000 pages of transcriptions and notes—all of which are scheduled for digitization.

Don Yoder Pennsylvania German Collection. In the 1950s, folklorist Don Yoder used a wire recorder to document Pennsylvania German songs and narratives. This collection is made up of 32 wire spools. Transcriptions taken from these wires have been used in publications (see Boyer 1951; Buffington 1974; and Yoder 1961), but the sound recordings have never been accessible to researchers.

Eleanor Dickinson Collection. Eleanor Dickinson researched the Holiness and Pentecostal churches of Appalachia and in the process made 169 black and white, open-reel video recordings of church services, tent meetings, interviews, and other aspects of mountain religion (Dickinson 1974; Maguire 1981). These videos are part of the Save Our Sounds Project, but not her 200 audio tapes and several hundred manuscript pages.

Zuni Storytelling Collection. This collection consists of 222 audio tapes. Recorded in 1966 and 1967 in Zuni Pueblo, New Mexico, 19 Zuni elders tell over 800 stories, including seven or eight narrators relating hour-long *telapna:we*, a traditional form of Zuni folktale (for similar material, see Tedlock 1999).

International Storytelling Foundation Collection. This organization is responsible for the annual National Storytelling Festival, as well as other public events. The collection comprises a comprehensive documentary record of every year of the Jonesborough, Tennessee, festival that began in 1973 (National Association for the Preservation and Perpetuation of Storytelling 1991; Smith 2001). This collection consists of 5,221 audiotapes and DAT tapes, 1,161 videotapes, 27 CDs, 174 LP discs, 1,200 volumes of books, 18 binders of the serial *Yarnspinner*, and approximately 196,000 manuscript leaves. The Save Our Sounds project will digitize all of the 678 open-reel audiotapes and 400 DAT tapes, as well as any manuscript documents directly related to these tapes.

Pearl Harbor Collection. As explained above, this collection has special political significance. Following the December 7, 1941 attack on Pearl Harbor and the subsequent declaration of war, the Library of Congress organized "man on the street" interviews around the country to document people's reactions to these events. Alan Lomax and other experienced fieldworkers conducted interviews in Washington, D.C., Tennessee, New York City, and Texas, among other locations, on December 8–10, 1941, and again in January and February 1942. A number of these discs were used for radio programs during World War II (Gevinson 2002). The collection contains 77 acetate discs and 90 pages of manuscript material, all of which are part of the Save Our Sounds Project.

Digitizing the sound recordings in these collections follows a strategy worked out in consultation with the Motion Picture, Broadcasting, and Recorded Sound Division of the Library of Congress. Audio engineers will first examine the physical condition of the individual recordings and perform any necessary cleaning of surfaces and grooves. This is the only type of cleaning done to recordings, which are otherwise recorded flat without any attempt to clean them electronically or enhance their sound. In this way, a recording is treated as an artifact—cracks, clicks, and all—with as much of its recorded information as possible available in digital form. Using the same philosophy, discs are recorded in stereo, even though they are mostly monographic recordings, since slightly different information might be found on one groove wall as opposed to the other.

Each recording is transferred to three digital files: one preservation master and two service copies. The preservation master is transferred at 96 kHz/24-bit word length as a WAVE file. The high service copy is a 44.1 kHz/16-bit "CD quality" WAVE file; and the low service copy is an MP3 file. All of these files are stored on Library of Congress servers and accessed from them. In addition, where no analog preservation master currently exists for a recording, we are transferring the item to 1/4-inch audio tape on 10-inch, slotless, NAB hub reels.

This digitizing strategy should allow the permanent storage of the recorded sound in a system where the digital file can be continually migrated to ever-newer hardware and software without deterioration. Of course, the size and capabilities of the Library of Congress allow for this system of server storage. Smaller repositories will probably have to rely on CDs or other physical data-storage formats that might, over time, also degrade or deteriorate.

There are exceptions to this strategy for certain types of formats. For example, there is no use creating a 96/24 master of a DAT tape, which is itself a digital medium; rather, a 44.1/16 WAVE file serves as a master, while the MP3 copy is the service file. The Dickinson video tapes are also an exception. Because of the excessive amount of storage required for digital moving image files, Dickinson's open-reel videos have been transferred to analog BetaSP cassettes and digitial DigiBeta cassettes as preservation masters. The service file is an MPEG3 streaming video on the library's servers.

Because we conceive of a sound recording as a package of information, the Save Our Sounds Project also digitizes the following material that accompanies the sound recording.

- ▸ The cover or housing, if it contains substantial information about the recording—field notes written on disc sleeves and recording logs on tape boxes being two examples.
- ▸ An image of the recording itself, if it is of interest because of its deteriorated or broken state.
- ▸ Accompanying notes, such as paper log slips inserted in cylinders or pages of notes kept inside of tape boxes.

Completing the concept of a sound recording as a package of information, each item receives extensive metadata description. Carl Fleischhauer will explain in greater detail the metadata standards used, both in the Save Our Sounds Project and as part of the digitizing strategy of the Library's Motion Picture, Broadcasting, and Recorded Sound Division. Using a Web-based Oracle database, each recording receives a metadata component of descriptive, administrative, and technical information. The master and service digital copies of the recording, images of the recording, its housing, or accompanying material are associated with this data to create the entire information package.

The result of this digitizing procedure is a virtual presentation of the collection. Researchers will be able to gain access to the collection in the American Folklife Center reading room, where they can listen to a recording, see associated images, read associated texts, and see all of the metadata related to the recording.

The ultimate goal, of course, is to make these collections available to everyone over the Internet. This goal has already been achieved in the case of one of the Save Our Sounds collections—the Pearl Harbor Collection. Because it is a relatively small collection that was originally generated by the Library of Congress, there were few problems in exhibiting the digitized sounds and manuscripts as one of the Library of Congress's American Memory sites (American Folklife Center 2003). These sites present significant bodies of material from different Library divisions, including collections from the American Folklife Center. In the case of the Pearl Harbor Collection, we were able to use the American Memory site to present a complete collection—in fact, three complete collections that make up the Pearl Harbor material held by the Archive of Folk Culture.

As other collections in the Save Our Sounds Project become available in digital form, they will be considered for some form of Web site presentation. Some, such as the Zuni Storytelling Collection, will probably remain restricted to the library's reading rooms, given the culturally sensitive nature of the narratives and performances in the collection. The same may be true of the Eleanor Dickinson Collection of religious practices.

The great song collections of James Madison Carpenter and Eloise Hubbard Linscott, however, will undoubtedly become available through a library Web site. Presenting these collections in such a way constitutes a form of mass media broadcasting, which involves at least one more step in the process of making these collections accessible. Because early collectors of folklore never sought release forms from those they recorded, the library is under the ethical and perhaps legal obligation to make a "good faith effort" to contact the original performers or their descendants to gain permission to broadcast their performances. This final step in the digital presentation of ethnographic material brings the American Folklife Center back to its core activity of involving tradition-bearers in building Library of Congress collections.

Bringing this project to fruition involves a great many players: the directors of the two centers at the Library of Congess and the Smithsonian Institution who, with their staff, developed the project, applied to the National Park Service, and who have continued to campaign for matching funds from outside donors; the donors themselves, who range from individuals to companies (such as Emtec Pro Media and the A&E History Channel) to foundations (such as the Grammy Foundation and the Rockefeller Foundation). The Leadership Committee of the Save Our Sounds Project has also played a role—especially through its Chairman, Mickey Hart—in finding support for the project. As well, many divisions of the library have been involved in the project (such as the Motion Picture, Broadcasting, and Recorded Sound Division; Financial Services; Office of the General Counsel; the National Digital Library; Contracts and Logistics Services; Automation Planning; and Information Technology Services); and outside contractors (such as the Cutting Corporation, UTA, and VidiPax) have been instrumental in digitization and metadata structure.

Librarians, archivists, sound engineers, information specialists, and other professionals have also assisted this project, either indirectly through their writings and other communications, or directly through the advice they have given and the questions they have asked. Various workers on the Save Our Sounds Project have attended meetings and workshops—such as this one—in order to share their experiences and learn from the work of others. At present, there are no national or international standards for the digital preservation of multi-media, ethnographic archival collections. Large centers such as the Library of Congress and the Smithsonian Institution will undoubtedly play an important part in establishing such standards, but only in conjunction with other centers—large and small—who hold similar kinds of material.

The responsibility, therefore, of librarians is to keep lines of communication open, and to strive for systems and procedures that can be shared with or replicated at other centers. Not only will such openness prevent the re-invention of the wheel (which has already happened to some extent), but will facilitate the eventual linking of digitized collections among institutions, or even the sharing of sites and data among institutions.

The time frame for the Save Our Sounds Project extends from June 2000 to September 2004, at which time all of the 3,000 earmarked recordings at the American Folklife Center will have been digitized and made accessible to researchers. But the result of this project will extend beyond the digitization of this group of recordings. The practices and procedures developed through this project will become the benchmark for the further digitization of the holdings of the Archive of Folk

Culture. Ultimately, the project will function, for better or worse, as a model of how ethnographic collections were digitally preserved at the beginning of the 21st century.

References

American Folklife Center, Library of Congress. *After the Day of Infamy: "Man-on-the-Street" Interviews Following the Attack on Pearl Harbor.* American Memory: Historical Collections for the National Digital Library. June 6, 2003. http://memory.loc.gov/ammem/afcphhtml/afcphhome.html.

Baker, Holly Cutting. "The Linscott Collection." *Folklife Center News* 3, no. iii (1979): 6–7.

Bishop, Julia C. "'The Most Valuable Collection of Child Ballads with Tunes Ever Published': The Unfinished Work of James Madison Carpenter." In *Ballads into Books: The Legacies of Francis James Child*, ed. T. Cheesman and S. Rieuwerts, 81–94. Bern, Switzerland: Peter Lang, 1999.

Bishop, Julia C., David Atkinson, Elaine Bradtke, Eddie Cass, Thomas A. McKean, and Robert Young Walser, eds. *The James Madison Carpenter Collection Online Catalogue*. 2003. http://www.hrionline.ac.uk/carpenter/index.html.

Boyer, Walter E., ed. *Songs Along the Mahantongo: Pennsylvania Dutch Folksongs.* Lancaster, PA: Pennsylvania Dutch Folklore Center, 1951.

Buffington, Albert F., comp. *Pennsylvania German Secular Folksongs.* Breinigsville, PA: Pennsylvania German Society, 1974.

"Dialect Collection for Folk Archive." *Folklife Center News* 8, no. 2 (1985): 4–6.

Dickinson, Eleanor. *Revival.* New York: Harper & Row, 1974.

Gevinson, Alan. "'What the Neighbors Say': The Radio Research Project of the Library of Congress." In *Performing Arts: Broadcasting*, 94–121. Washington, D.C.: Library of Congress, 2002.

Gray, Judith A. and Dorothy Sara Lee, eds. *The Federal Cylinder Project: A Guide to Field Cylinder Collections in Federal Agencies.* Volume 2: *Northeastern Indian Catalog*; *Southeastern Indian Catalog.* Washington, D.C.: American Folklife Center, Library of Congress, 1985.

Fewkes, Jesse Walter. "A Contribution to Pasamaquoddy Folk-Lore." *Journal of American Folklore* 3 (1890): 257–80.

Kurath, Hans, ed. *Linguistic Atlas of New England*. Providence: Brown University, 1939–43.

Linscott, Eloise Hubbard, ed. *Folk Songs of Old New England*. New York: Macmillan, 1939.

Maguire, Marsha. "Confirming the Word: Snake-Handling Sects in Southern Appalachia." *The Quarterly Journal of the Library of Congress* 38 (1981):166–79.

National Association for the Preservation and Perpetuation of Storytelling. *Best-Loved Stories Told at the National Storytelling Festival*. Jonesborough, TN: National Storytelling Press; Little Rock: August House, 1991.

National Park Service and National Trust for Historic Preservation. *Save America's Treasures*. n.d. http://www.saveamericastreasures.org/.

Smith, Jimmy Neil. "Storytelling Collection Comes to the Library of Congress." *Folklife Center News* 23, no. 3 (2001): 3–5.

Tedlock, Dennis, trans. *Finding the Center: The Art of the Zuni Storyteller*. 2nd ed. Lincoln: University of Nebraska Press, 1999.

Yoder, Don. *Pennsylvania Spirituals*. Lancaster, PA: Pennsylvania Folklife Society, 1961.

The Case for Audio Preservation

Karl Miller
Lecturer, Preservation and Conservation Studies
School of Information
University of Texas at Austin

THE CHOICE TO ADDRESS AUDIO PRESERVATION, and preservation in general, may well reflect an organization's perspective on the future of libraries and archives. In an evolving environment, new operational modalities will be required even if the financial base remains constant. In the for-profit environment, new operational modalities can increase productivity as in the development of a new product line. The benefits (profit) in the non-profit sector are not as easy to measure. It is difficult enough for any administration to balance the exegesis of one function over another, let alone to address the introduction of a heretofore-unaddressed activity.

Few would argue that libraries and archives have not experienced a period of change in recent years. These changes have made many question the very relevance of libraries. Over twenty years ago James Thompson wrote, "In terms of size, arrangement, and catalogues, the conventional library has reached an organizational and financial impasse. Coincidentally there has emerged a preemptive new technology for the storage, handling, and transmission of information, potentially better suited to the convenience of users. Libraries may disappear like the dinosaurs; or they may, by returning to first principles, be able to adapt and successfully survive."[1] As to what those "first principles" might be, Thompson continues, "to provide online access to resources for individuals who do not have their own terminals; libraries to be centres in which trained personnel will be available to assist the user to exploit databases and databanks; libraries to take chief responsibility for materials of purely local and very specialized interest; libraries to provide a high-quality question-answering service based on a vast shared 'electronic encyclopaedia'; and libraries to act as centres for important community information services."[2]

Clearly with the advent of low cost computers and wireless communications, the provision of workstations may not be relevant to libraries of the future. While libraries certainly have "trained personnel," the skill sets have been changing. In many respects, libraries have deemphasized some aspects of their operations, which could be seen as core strengths. For example, with the advent of the outsourcing of selection via approval plans, the notion of the subject specialist, conversant in the vernacular of a discipline, is increasingly becoming a rarity. While there are questions of the quality of much of the information published on the Web, it is not uncommon for an individual to complete an undergraduate degree without ever having been in a library or relied upon those "trained personnel."

As for that "preemptive new technology for the storage, handling, and transmission of information," the bulk of information is being created with that new technology, the computer. "The world produces between 1 and 2 exabytes of unique information per year, which is roughly 250 megabytes for every man, woman, and child on earth. An exabyte is a billion gigabytes, or 1018 bytes. Printed documents of all kinds comprise only .003% of the total. Magnetic storage is by far the largest medium for storing information and is the most rapidly growing, with shipped hard drive capacity doubling every year. Magnetic storage is rapidly becoming the universal medium for information storage."[3] Relative to the storage and retrieval of information, print is ephemeral, yet it remains central to the acquisitions budgets of most libraries. It would seem, based on casual observation, that libraries are having a difficult time defining their function.

A notion understood by archives but not often given a high priority in libraries can be found in the subheading of "Ethical Considerations," in the technical committee report of the International Association of Archives: "There are four basic tasks that are performed by all archives—acquisition, documentation, access, preservation. The primary task, however, is to preserve the information placed in the care of the collection."[4]

If one accepts some of the notions expressed by Thompson, the future of libraries perhaps resides in the core strengths of its own past as well as in the artifacts of the past, and in the unique materials placed in its trust, those "materials of purely local and very specialized interest." As for the activities and services related to those materials that can cannot reasonably be expected to be automated, one could look to the preservation of unique materials and informed access to those collections, areas historically undervalued in libraries, but central to archives. Unique recordings are amongst the most significant artifacts of the past. To minimize the value of audio recordings is to minimize the importance of the information we receive through our sense of hearing.

Economics of Audio Preservation

Staff. Audio preservation is problematic for a variety of reasons. Unlike print, recorded audio cannot be auditioned without playback equipment. It is also necessary to have the technical background and listening skills to know what represents optimal retrieval of that audio. Machines have to be maintained in proper calibration, and often times, such technicalities as equalization require specialized listening skills. Even many well-informed technicians and musicians can make plausible arguments for differing perspectives on such fundamental issues as pitch in a recording. What size needle is needed to provide the least amount of noise in playback and is this desirable as the greater the noise level, the easier it can be identified by much of the noise reduction software? These are but a few of the variables that indicate that there cannot be an absolute in reformatting. This leaves us with the primary and central expense to any audio preservation program: technical expertise coupled with a trained ear and an in-depth knowledge of preservation techniques. Knowledge of the subject and what recordings may or may not be unique (what deserves reformatting) is similarly requisite. It is rare to find all of the requisite skills in a single individual.

This one component, informed personnel, is perhaps the most expensive long-term investment, the most needed, and the most difficult to acquire. Unfortunately, at this time, there is no certification in audio preservation and only one institution of higher learning in this country is offering classes in audio preservation, namely, Preservation and Conservation Studies at The University of Texas at Austin.

Work Space. Audio exists in time. Digitizing analog audio requires real time playback and the full attention of the technician. This suggests the technician needs the opportunity to work uninterrupted in an acoustically isolated environment. Not only does the isolation provide the technician the opportunity to listen at amplitudes that are needed to adjust his or her work without disturbing adjacent work areas but, of equal importance, it keeps extraneous noise from disturbing the monitoring of the audio. Drop-in soundproof modules can be purchased for under $4,000.

Standards. There are no mutually agreed upon standards for audio storage. Any cost estimates will be subject to the level of resolution and target format for reformatting and your preservation of that target format. Even the fundamental question of what sampling rate should be used is debated. The audio CD features 44,100 samples per second with a resolution of 16 bits. Is this adequate? "To approximate a high-fidelity sound with a bandwidth of 15,000 Hz, we require 30,000 samples per second, or a sampling time of 1/30,000 per second."[5] Today our standards are higher. Digitizing a sine wave of 12,500 cycles per second, a pitch near or just beyond the top of the available range of hearing for those of us past the age of 50, will clearly demonstrate the limitations of a sampling rate of 44,100 per second. A graphic display of the wave will show significant distortion. The new audio standard calls for a resolution of 24 bits with 96,000 samples per second. Even higher resolution and sampling rates are considered desirable by some. Choosing an appropriate standard to accommodate the needs of a particular institution or group of materials is yet another aspect that requires informed staff assistance.

Then, what of the final storage format? "It must be stressed that coding schemes used for preservation purposes must be openly defined and not proprietary to a limited number of manufacturers."[6] Further, "it has become generally accepted that, when selecting a digital target format, formats employing data reduction (frequently mistakenly called data compression) based on perceptual coding (lossy codecs) must not be used."[7] Oddly enough the "aging" technology of the CD-R remains as the one universally accepted format which can accommodate these criteria.

Equipment. One of the often spoken axioms in audio preservation is "it is better to do something than nothing." A lacquer disc not transferred is likely to be lost. If your organization is not willing to devote the resources to "do it right," low cost equipment can handle some of the more familiar formats. There are inexpensive systems that combine cassette playback with a CD recorder in a single unit costing less than $700. That system also includes inputs for a magnetic cartridge. There are also stand-alone systems that will partially automate the digitization of reel-to-reel tapes.

While the primary equipment needs for an audio reformatting lab will be for the preparation of a "straight" or unaltered transfer of a recording, it is important that consideration be given to restoration. In many instances, perhaps due to an excessively high level of background noise, or a significant amplitude difference between an interviewer and his subject, a "straight" copy will be of little value to the patron. Therefore, it is recommended that any facility include equipment and/or software for restoration. A reasonable argument can be made for preparing a restored copy along with the "straight" copy as part of the normal process of reformatting. The noise reduction technology of today is affordable, and when used judiciously by a trained ear, can produce remarkable results. While there are always new technologies on the horizon (such as imaging for the playback of discs) which hold the promise for better noise reduction, the initial monitoring of the original during the transfer will identify items in need of restoration, making it more effective to do the restoration at that time. Considering the potentially large collections of recordings, it may be unreasonable to assume one will have time to deal with a recording more than once.

A reasonably well-equipped lab, designed for a moderate level of resolution, could easily consume $60,000 in equipment costs. This cost estimate includes basic analog playback of most disc formats, cassette tape and reel-to-reel, monitoring equipment, digital to analog converters, a computer for editing, basic restoration software, editing software, and CDR burning capabilities. For cylinder playback, one needs to add approximately $12,000. These costs do not include the recommended acoustically isolated studio or the cost of staff and supplies.

Outsourcing. Outsourcing audio reformatting can cost between $90 and $100 an hour. This modality still requires that the quality control be monitored and may present security problems for rare materials, which could need to be transported. An employee working a 40-hour week with a two-week vacation, holidays, and sick leave can be expected to be available approximately 1,920 hours a year. Subtracting time spent on equipment maintenance, meetings, participation in professional organizations, professional development, etc. many work models estimate productivity of professional positions at 80%. That could leave a full time employee with approximately 1,440 hours a year to do restoration. A salary of $40,000 equates to approximately $28 an hour. That does not include the cost of supplies, a work place, or the equipment. Outsourcing for 1,440 hours at $90 an hour could cost $129,600 versus $40,000 in house, leaving a balance in a single year of approximately $89,000 which could be used for supplies and equipment. The major benefits of outsourcing are for those organizations that have a finite quantity of audio to reformat or formats that require esoteric and generally unavailable analog equipment for playback.

Libraries frequently outsource the preservation of their digital information to computation vendors, database managers, and technicians not directly under the control of the library. However, it is essential that at least one informed technician be available on the library staff to create and monitor the specifications of any contractual agreements with outside vendors. The same is true for audio reformatting. Even if the bulk of the work is to be outsourced, it is essential that informed staff draft specifications and monitor the quality of the work. This requires, at the very least, the equipment and expertise to check what is on the recordings and to monitor the work done by the outside vendor. In-house expertise is also likely to be needed to provide and maintain patron playback facilities.

Cost Recovery. As with any expense, decisions regarding audio preservation and restoration should not be viewed in isolation, but considered in the light of the operational or systemic functionalism of the organization. Cost recovery can be measured in many ways. It can be viewed in the context of one task providing ancillary support to another, thus providing a reduction in expenditures. Similarly an institution that provides training in preservation or audio engineering might consider internships, directed at the processing of materials, as being beneficial to both the collection and the educational process.

The audio technician can serve as a resource to many preservation activities such as the determination of all digital storage modalities and the preservation of all digital information. This person can serve as a resource for the emerging preservation technologies. The experienced technician can provide mentoring of staff and assist in the preparation of materials for publication on the Web. There are also opportunities for lower level staff to perform reformatting operations, once an informed methodology is established, that can be monitored by a trained professional. When the copyrights allow, in rare circumstances, the publishing of donated recordings may help defray the costs of preservation.

The Future. Audio was the first information stored electromagnetically. In many ways, the study of the history of audio preservation is a guide for the future of the preservation of electromagnetic storage. It is a history filled with changing formats, lost information, all subject to quickly evolving market driven technologies. If the past is any indication, the future is not bright.

The uninformed often look to the digital technology as the savior of preservation. Unfortunately, "digital carriers must also be regarded as endangered by decay, especially if they have never been checked for their data integrity. Several formats already show obvious signs of chemical decomposition. Some are additionally threatened by obsolescence of hardware."[8] "To date, none of the digital recording systems developed specifically for audio has achieved a proven stability in the market place, let alone in an archive. The commercial lifetime of modern formats and systems is likely to become shorter and shorter. In the future this may increasingly lead to the obsolescence of hardware while the carriers are still in good condition…It is also likely that in some arenas there will be no physical carrier to distribute, in which event the issue of format obsolescence applies to the file format itself."[9]

Conclusion

Audio archives owe their existence to those librarians and archivists whose vision of the future embraces both the technology of the present and the future and the history of the past. It is for those who value the substantive information found in sound. When that sound is music, it can be what the great American composer William Schuman once said to me, "amongst the most noble expressions of the human spirit." It would seem when we lose any unique audio recording, we are losing a part of ourselves.

The ultimate solution to information preservation requires the development and/or ownership of proprietary software to access digital information, software that can be universally adopted, made available at low cost, and maintained indefinitely. It needs to be coupled with a storage technology with the permanence of stone. Only if such a utopian notion becomes a reality can we ever consider information preserved. As the quantity of information continues to grow exponentially, the knowledge of technology is a most obvious requisite skill. Less obvious a need, but of equal value, are those professionals with the subject expertise to make the informed decisions regarding what will be preserved. Ultimately, as the economic limitations and even the theoretical limits of information storage tell us, we can't save it all.

Endnotes

1. James Thompson, "The End of Libraries," *The Electronic Library* 1, no. 4 (October 1983): 245.

2. Ibid., 254.

3. http://www.sims.berkeley.edu/research/projects/how-much-info/summary.html, c.2000 Regents of the University of California.

4. International Association of Sound Archives, *Standards, Recommended Practices and Strategies, Version 2 (*September 2001*):* 3. Also available online: http://www.iasa-web.org/iasa0013.htm.

5. Max V. Mathews, *The Technology of Computer Music* (Cambridge: M.I.T. Press, 1969), 5.

6. *Standards, Recommended Practices and Strategies*, 7.

7. Ibid., 8.

8. Ibid., 11.

9. Ibid., 5.

Contracting for Services

A Dialogue between **Anji Cornette**, The Cutting Corporation,
and **Alan Lewis**, National Archives and Records Administration

LEWIS: Good morning. I'm Alan Lewis from the National Archives and Records Administration (NARA) in Washington. However, my participation here today is not as an official spokesperson for the Archives. Although I'm officially identified as an "Audiovisual Preservation Expert," I am not in the agency's Preservation Division but rather am an AV person in our machine-based AV curatorial unit. I've become a contract manager for out-of-house audio, film, and video reproduction work. Since much of what is being discussed in this conference deals with *preservation policy*, my views are my own and based on my experience both at NARA and elsewhere in the AV field.

My co-presenter this morning is Anji Cornette, who will introduce herself in a minute. Her company has been one of my contract laboratories and thus I have had a five-year relationship with her company. Our joint presentation builds not only on that relationship but also on our independent experiences at other times and places, with other laboratories, and with other institutions.

Indeed, government contract administrators strive to an arms-length distance between themselves and their contractors. However, in the real world, once a contract is in place, both parties, if they are intelligent about it, can and should work toward a professional win-win relationship.

CORNETTE: Hi, I am Anji Cornette, Division Director of The Cutting Corporation. The Cutting Corporation has been in audio production for over thirty years and specifically working in sound preservation for over twenty-two years. I have been with the company for fourteen years. In addition to sound preservation, our company is very active in setting standards. I serve as co-chair of the R6 Mobile Electronics, WG11 Spoken Word Committee of the Consumer Electronics Association, which is currently working on standards for the download delivery of digital audio files.

LEWIS: In my experience, a collection manager may have need for two kinds of contract services: laboratory reproduction services and off-site storage services. In our all-too-brief time with you, Anji and I can cover only the first by:

- ▸ Defining the needs for laboratory services.
- ▸ Talking a bit about laboratory services themselves.
- ▸ Discussing conservation versus preservation versus restoration.
- ▸ Creating a statement of work by which to solicit and manage a contract.

Turning to the first topic, **Defining the Need**, I think there are three reasons to do duplication work:

- ▸ **Failure of the medium.** That's the stuff on the shelf that is dying because of its own inherent vice (as the archivists call it) or because of the damaging things we do to it or allow to happen to it.
- ▸ **Technical obsolescence.** That's the machinery needed to playback the media, the replacement parts to keep them going and the knowledgeable people to run and maintain them.
- ▸ **Researcher need for access.** That's all about not allowing your best preservation copy to be used by every researcher who comes through the door because every time a piece of recorded media is put on a machine there's the potential or the fact that damage will occur to it.

Concerning failure of media, during the course of the century plus of recorded sound history, many, many materials have been used as a base material, a binder, or as a surface coating on audio recordings. They include aluminum, Bakelite, cardboard, cellulose (cellulose acetate), celluloid (cellulose nitrate), ferrous metal, glass, paper, plaster of Paris, PVC, rubber ("Vulcanite"), shellac, styrene, wood products, other plastics, wax, etc. Some may have been ideal for the purpose to which they were put, initial recording or distribution for a relatively short time, but most have drawbacks that preservation laboratories must be geared up to deal with. Anji has prepared a series of images documenting some of the challenges her lab has seen recently.

CORNETTE: As a vendor with numerous years of experience in sound preservation, The Cutting Corporation has seen many institutions begin to notice deterioration of formats in their sound collections and therefore have come to us. The problems range from sticky shed syndrome to vinegar syndrome, to broken discs, to discs that have lacquer peeling off. There are discs in collections with issues such as palmitic acid or powder residue. Some reel-to-reel tapes have deposits of dirt and mold. We were even asked to consult with an institution in the Caribbean that had a collection suffering from biological infestation. More on that subject later.

The earliest commercial recording media—wax cylinders—is where we might start. The raw material that typically composed a wax cylinder was vegetable wax. For example, Edison solid wax cylinders were composed of ceresin wax, beeswax, and steric acid. Because of this, it is in the nature of wax cylinders to be fragile and breakable. They are also prone to mold and fungal growth especially in warm, moist, and dark environments. Mold damage is due to high relative humidity (RH) that is created by the individual cylinder packaged in hygroscopic cotton or wool wadding. Cylinders can also be attacked by fungus and the residue is fungal mycelium or animal bacteria that eats the wax. The fungus feeds on the surface of the cylinders and the audio program can be lost. Some waxes had oxides and oils that surfaced as white or blue haze on the cylinders and sometimes were mistaken for mold.

Moving to *first generation* or *instantaneous discs*, these so-called *acetates* were manufactured with an aluminum, glass, or cardboard base. The base was then coated with nitrocellulose lacquer plasticized with castor oil. This was an unstable mixture making these acetates not suitable for long-term storage. Symptoms or problems include continuous shrinking of the lacquer top coating, embrittlement, and irreversible loss of recorded sound because of the loss of the castor oil plasticizer. Since the core does not shrink and the lacquer coating does (or expands under changes in temperature) cracking and peeling of the lacquer coating results.

The production of palmitic acid is caused by the hydrolysis of the castor oil from heat and humidity, which then oozes through the lacquer on a disc. The specks or small mounds on the groove look similar to powder residue but have a more crystallized appearance. Palmitic acid is stubborn to remove and requires extensive cleaning. Powder residue may appear on lacquer discs as dried white specks or pasty mounds on the grooves. The main symptom is caused by glue from the paper label, which has spread over time onto the recording surface of the disc. Sometimes powder residue is mistaken for mold or palmitic acid.

Two years ago, The Cutting Corporation retrieved a collection of 1,300 recorded discs. The archivist thought the records were all in good to fair condition. In actuality, when we retrieved the collection, we found that two thirds of the discs were in poor condition suffering from both powder residue and palmitic acid. Each disc required extensive cleaning.

Before WWII, wire recordings were made with steel wire, which can rust but can easily be cured by wiping them down unless the rust is severe. On magnetic wire recordings, print through is a problem. But the worst problem for wire recordings is what we call the bird's nest syndrome or tangles and snarls that lead to the wire breaking.

Wire recordings existed in parallel to 1/4-inch audiotape and eventually were replaced by tape. Cellulose acetate reel-to-reel tapes and acetate discs are subject to a slow form of chemical deterioration known as *vinegar syndrome*. The main symptoms of this problem are a vinegar-like odor and buckling, shrinking, and embrittlement of the tape or cellulose disc. Low temperature storage conditions can aid in slowing down this process.

A well-known institution sent us two collections, one with a big surprise. When we opened the box, it was like sticking your head in a bag of salt and vinegar potato chips. The collection to our surprise was afflicted with vinegar syndrome. More on vinegar syndrome can be found in an article, "Vinegar Syndrome: An Experience with the Silent but Stinky Acetate Tape Killer" published at http://www.cuttingarchives.com. There is a product by the Image Permanence Institute for measuring the level of acidity. The color on the strip, blue, will change to yellow as it detects high acidity, which is what a mustard yellow color on the strip represents.

Polyester magnetic tape stock that came into wide use in the 1960s can develop a condition known as sticky-shed syndrome. This problem occurs when oxidation of the tape sticks to the guides and magnetic heads of the playback machine. The material builds up a residue on the guides and heads as playback continues. This causes distortion to the sound of the recording that is called separation or shed loss. It results in very low-level volume, fuzzy sound, or inaudible audio. We have an institution that thought their reel-to-reels were in good condition. The reels did not display any sign of deterioration and had been kept in a climate-controlled environment. When we actually started to make the preservation transfers, we found to our surprise that the reels were affected by sticky-shed syndrome. In some cases, the sticky-shed was so intense that the reels required double or triple baking before a transfer could be made.

Here's an example of what a sticky-shed tape sounds like. (Played sample of tape with sticky-shed syndrome.) Hydrolysis is the process by which moisture is absorbed by a material. With magnetic tape, it is caused by extreme humidity that results in the magnetic tape binder weakening and the binder with its oxide information carrier peeling or dropping off the tape. The results are dropouts, shedding, or complete detachment from the base.

Mold is caused by the growth of fungus in elevated temperature or humidity conditions. It can cause serious distortion and physical breakdown in most audio formats, both grooved and magnetic formats alike. The other major agent involved in fungal action is the presence of organic material on the recording medium due to unclean storage areas. We received broadcast tapes from an institution that had been stored for years in a damp and moldy basement. The tapes had tiny specks of mold on them that had to be delicately removed in order to get a clean transfer.

On occasion, due to the environment in which the audio materials are stored, the materials will be affected by unusual conditions such as biological infestation. Due to hot and humid conditions and a basement flood where some reel-to-reel recordings were stored, termites decided that they would make their home in the reels. They ate through the reel boxes and resided on the tape under the plastic reels.

Belt recordings, another technology and one often used for office dictation, suffer from severe creases because it is not unusual to find them being stored flat in file folders along with copies of the letters or reports that were typed from them. Belts were often marked with crayons or wax pencils to denote the beginning and end of letters.

And then we come to the Memovox disc, a cellulose acetate grooved recording medium that is prone to the *sombrero effect*, in which the edges take on a scalloped shape. This is especially true if they were stored vertically in a box. In order to transfer the disc, it has to be flattened as much as possible.

LEWIS: Having made the case that there are many recording media used over these past 100-plus years and there are problems with them, we also recognize that the technologies have come and gone. The fact that so many have gone gives rise to reformatting in laboratories in order to rescue the content.

First there were the cylinders, the first of the physical forms, both in their one-off original copies and later mass-produced ones. Then came flat discs using mechanical stylus-in-groove vibration-capture technology, in one form or another, that lasted into the 1990s. Now the new disc formats use laser technology. In parallel with discs, ribbon-like media were developed in conjunction with magnetic recording systems. The first used paper tape with a magnetically sensitive coating and then metal bands like bandsaw blades. At about the same time, PVC flexible tape with the magnetic material suspended in the substrate itself as developed.

Recording on wire—another linear medium—was developed, more-or-less successfully. Later, taking a page from motion picture film technology, cellulose acetate-base tape replaced paper-base tape until it too gave way to polyester-base tape. So the question and the challenge to archives and reformatting laboratories becomes, "What do you do when you have media and no machines?"

CORNETTE: As a vendor, one of the exciting aspects of sound preservation is dealing with obsolete recording technology. Playback machines are not necessarily available and each medium has its own challenges. The Cutting Corporation has had to either find and fix an obsolete machine or reverse engineer an obsolete machine usually cannibalizing parts of other machines, to recreate an older technology. We have conducted research to locate existing machines and if those results do not yield anything, then we rebuild or reverse engineer.

Over the past few years, we have had to rebuild a Memovox machine and we had to reverse engineer a Dictalog Magnabelt machine. Fortunately, we had extra parts from other disc and belts technologies so we were able to cannibalize the parts in order to build the replacement machines. There's a lot of trial and error before we get the machines working correctly, and it takes hours and hours of engineering time, patience, and ingenuity. There are many intricate parts that have to work together to get the speed right. For our Dictalog Magnabelt machine we had to deal with a belt of a different length and width than Dictabelts or Magnabelts. Our engineers succeeded by building a machine with a lengthened and wider belt path and a longer lead screw to move the playback head across the wider belt.

Also in the realm of machinery, custom-built devices may be needed to solve the cleaning and handling of items before preservation. This is The Cutting Corporation's proprietary Open Reel Vacuum Cleaning System to clean the front and back surfaces of a tape. Such machines must clean delicately but effectively without hurting the media. To do it, the lab's technical engineers worked closely with the sound preservation engineers to modify a reel-to-reel machine into an audiotape cleaner. The two engineers had lengthy discussions and eventually took the heads off the machine, added 1/2-inch Tape Wipe and a low-pressure vacuum cleaner. The reels are cleaned of dust and dirt and the sound transfer engineers are now able to get clean playback of dirty tapes in order to create a new preservation masters.

LEWIS: The final need for reformatting—or perhaps in this case just duplication—is serving the needs of researchers. Having material on the shelf in good storage conditions is only half the archival task. Access is the other.

It is not sound (pardon the pun) archival practice to allow researchers to use the sole, best quality and likely irreplaceable archival copy of some unique record. Hence, a surrogate copy, a service copy, a reference copy—whatever you might call it in your shop—needs to be made from the original. To my mind, it should reflect the sound of the original recording with only minimal *signal processing*, if any, done to it so that a researcher hears the program content more or less as the recording exists today or perhaps just as a listener would have heard it at the time the recording was new. Whether you allow minimal signal processing or prefer your access copies to be wholly unprocessed, a *flat transfer* as some call it, is a matter for your archive to decide.

Also in practical terms, other users may have more than academic needs. They are the *repurposers*, the people who make new *product* out of old. They're the ones who should be paying the extra costs of making a Caruso cylinder or disc sound like it was recorded yesterday. (I won't get into the ethics of this whole matter of *improving* historic recordings.)

CORNETTE: Once preservation masters have been made, institutions often order service copies or reference copies for researcher access. In the past, many institutions asked for analog reel-to-reel tapes or cassettes but over the last few years, there has been a gradual shift to digital formats such as CD-Rs for reference copies. Other formats requested have been digital formats such as MP3 files for download or Real Audio Streaming files. One institution was resistant to having digital service copies because their parent institution was sticking with analog media. After two years of resistance, the institution decided to have us make CD-R reference copies in addition to the analog reference copies and found that the CD-R reference copies sounded very good and were convenient to use. Ever since, they have ordered CD-R reference or service copies instead of analog tape

LEWIS: Having defined the need for lab work, please note that a media preservation laboratory is not a run-of-the-mill audio production house, the one down the street or around the corner run by a rock-and-roller whose hearing is something less than wonderful. It must be an appropriate physical plant, equipped with the right equipment and staffed by the right people who are sensitive to and experienced with old media and who are clear about three services they might provide you. Those services are conservation, preservation, and restoration.

I think of conservation of an original item as providing a number of services. Among them are doing a physical inspection and preparing a condition report, doing proper winding of linear media like tapes, making repairs, and/or cleaning the item, and rehousing the item in preparation for long-term storage.

To my way of thinking, preservation deals with saving the recorded content of the original item: inspection of the original first, cleaning and providing heroic measures if the item is deteriorated, and duplication of it to a current format or technology, and perhaps making duplicating masters and access copies at the same time. If the original item is so physically troubled or its technology is so exotic, this is the time that a *Replacement Preservation Copy* would be made.

Finally, restoration might mean trying to ferret out what the original recording was supposed to sound like at the time period in which it was made and replicating that. Or maybe, it is making an old recording sound like it was made yesterday. The bottom line is that you and the laboratory need to decide *how much restoration is restoration* if you are going to do any restoration at all.

To recap a bit, Anji and I have talked about laboratory services, perhaps the major AV archives service that is contracted out. Those services are driven by:

- ► Failing media—and triage the collection first in order to do the most important failing recordings first.
- ► Obsolescence of the technology—and triage the collection to do the most important obsolete recordings first.
- ► Reproduction demands for research or repurposing.

The second contract service I would have talked about is off-site storage services—but because of time limitations and the fact that I covered some of this somewhat obliquely in my session yesterday, we'll save this for the next Sound Savings Symposium.

Now on to the Statement of Work or the SOW as we sometimes call it. It is important for you, the client, to have in mind what the project is all about. This is the point where the vague yearnings that "something needs to be done" must move to the next step: defining what it is that has to be done. The goal statement should be written, in my estimation, in order to help you and prospective vendors focus on it.

CORNETTE: As vendors, it is helpful when the archivist knows the history of the collection. What is the subject matter? When was the collection originally recorded? Under what conditions were the recordings stored before the archivist received them? It was helpful to us that the Smithsonian knew that J.P. Harrington's recordings slowed down because the battery for his recording machine was dying. A U.S. Department of State employee brought back a gift that a Russian friend had given him of a record that was an actual x-ray with embossed grooves on the x-ray. He knew that the record was made during the Stalin reign but did not know what was on the record. It ended up being a recording of Elvis Presley singing a song from one of his upcoming movies. The x-ray record played beautifully on our turntable with a 2.7 mil stylus.

Some recordings we have transferred are rich in American history. For example, we have listened to recordings Jane Fonda made during the Vietnam War. We listened to Sergeant Tom as he defected during the Vietnam War and all the propaganda he created. We have recovered a daughter's memory of her deceased mother's voice. We have listened to recordings of Nobel Prize physicists. We have listened to Duke Ellington and Ella Fitzgerald in jam sessions and rehearsals with their friends and colleagues. We have gotten to know Brownie Wise, the Tupperware Lady, and reminisce with Hills Brother Coffee commercials and Eskimo Pie commercials.

LEWIS: After having gone through the process of deciding what needs to be done, this is the step in which you really get down to nuts and bolts of defining just what's on the shelf that has to be dealt with. A clear inventory is needed for the vendor to understand the amount of work to be done and to bid the job properly.

CORNETTE: As a vendor, it helps when the client or institution knows as much information about their sound collections as possible. In the past, we have had institutions call up at the time that they are ready to go after a grant to find that they really don't know what's in the collection. Often this makes it extremely hard for us to provide a ballpark cost estimate. Collections can have reels with different speeds, more than one track, different thickness, and deterioration issues such as sticky-shed syndrome and vinegar syndrome. Some collections may have been badly packed or tight-wound and therefore may have developed issues. Others may have mold and dirt and require extensive cleaning.

The same goes for discs. There are acetate discs and vinyl discs. The acetate discs can be based on glass, cardboard, or aluminum. They also can have a host of issues such as chipping, flaking, powder residue, and palmitic acid. Records also play at varying speeds, although most acetates are at 78 rpm. It helps when an archivist knows these types of information at the time of the bid process or cost proposal.

LEWIS: In this step, the end product of the project is defined. This is the step of getting beyond the stage of vague yearning to the substance of what needs to be done.

CORNETTE: A vendor needs to know what medium(s) the client wants to preserve to. Are they open to digital preservation or do they want to stick with traditional analog preservation and analog formats such as reel-to-reel tape? If they are open to digital preservation, are they willing to experiment with high digital compression formats? Do they have digital storage solutions at their institution? Would they like service or researcher copies for access purposes? Would they like a 44.1 kHz, 16-bit WAVE file on CD-R, or some other digital format like MP3? Do they have a database they would like the vendor to work with or would they like the vendor to provide a database? What information would they like on the labels of the preservation masters and the service/researcher copies? In some cases, institutions like the vendor to consult with them to provide suggestions for their end solutions.

The laboratory also needs to make sure the customer understands the difference between conservation, preservation, and restoration. There are institutions that have very small budgets and therefore have to conserve their collections by re-housing or tight-winding their materials. Customers can confuse preservation with restoration. In fact, recently a woman had an album of Yale University's exclusive musical group, The Whiffenpoofs, which she wanted us to preserve for her husband. The engineer preserved the album and then made a reference copy on CD with the tracks broken up by the bands in the record. We found out later that what the client actually wanted us to do was some remastering and break the bands into individual tracks of songs. She did not understand the difference between preservation and re-mastering. We have also had a customer ask for preservation and when they listened to the completed projected, they called to complain because what they really wanted was restoration or cleaned up sound.

LEWIS: In thinking through the project, this is where reasonable, realistic, and achievable landmarks are set to measure the productivity of the project.

CORNETTE: The laboratory should work with the vendor to decide start and due dates. There are also issues of picking up originals and delivering them. Does the client have requirements for picking up originals and returning them? How many pieces of the collection can go over to the vendor's facility at one time? Will it come in batches or in an entire collection? Does the client have specifications for storage and handling while originals are in the laboratory's facilities? The laboratory should follow strict measures for storage and handling of originals while in the vendor's facilities.

LEWIS: If you are a nonprofit or a government agency, you may be able to purchase raw stock for the new products or packaging materials at a lesser price. Will the reproduction vendor allow it?

CORNETTE: The laboratory should coordinate with the client on any outstanding items such as labels. The laboratory also needs to arrange for shipping. Once the lab knows what the approved materials are for the project, the laboratory should arrange with suppliers to get items in bulk for discount prices, which then can be passed on to the customer.

LEWIS: As a purchaser of reproduction services what are you seeking in a vendor for your precious, one-of-a-kind, archival original materials? If you think I'm preju-dicing you against the audio recording studio down the block—you may be right! Among the things I'd like to know are the company's length of time in business, its expertise in the specific tasks that will be involved in my project, the qualifications of staff who will have their hands on my materials, lists of equipment that will be used and how—and how often—the equipment will be cleaned, aligned, or otherwise maintained.

CORNETTE: The laboratory should have extensive experience in the area of sound preservation and sound restoration from working on a variety of sound collections for various institutions. The laboratory should have standard audio recording equipment as well as obsolete playback machines. The laboratory should have an impressive selection of styli and a variety of reel-to-reel record and playback machines. The equipment should be cleaned, aligned, and maintained on a regular basis. This requires having proper testing equipment at the laboratory's facilities and well-trained personnel including an outgoing quality control department. Engineers should have a degree or certificate in sound engineering and experience in some aspect of sound preservation both in the analog and digital realm. Some of the vendor's employees should have security clearance and the backgrounds of any engineers should be checked, as some of the material could be sensitive or classified.

LEWIS: Here's an interesting dilemma. You've got an original recording that you may not be able to play back because you don't have the equipment or it isn't quite gentle enough for that archival original, so how will you know the vendor has done the best job possible when you can't listen to them side-by-side?

CORNETTE: Quality assurance is conducted by the vendor before the preserved material goes to the client who often has a quality assurance program of its own set in place. The laboratory should adhere to a strict quality assurance program. The engineer should A/B the original to what he/she is preserving as a reference point before completing the entire batch. Once a batch is completed, it goes to the quality assurance department for a quality review. Additionally, before a project is sent on to the client, the engineering staff or manager should conduct an outgoing quality inspection. On occasion, the client and vendor may agree on something that might be redone differently, not necessarily because it was done wrong, but to capture the program in a different way. This is called rework and is conducted immediately to complete a batch.

LEWIS: As to labeling, how will the newly made recording medium be labeled as well as its container? With CDs and DVDs, what is safe? As to packaging, what's safe and protective? Are there institutional or professional standards? Color coding? Corporate logos? May the laboratory's name and address appear on the label?

CORNETTE: The vendor goes over with the client how the labels for the preservation masters and the service copies are to be done. An institution logo can be added on. Questions on color of label, any standard numbering for items, and any other information that needs to be on the label are answered and clarified before the vendor generates labels. The vendor should discuss with the client whether to use direct imprint or paper labels on the CDs. Reels should always be returned to the customer in a tight-wound position. The preservation masters should be housed in archivally approved containers to be housed at the client's facility. Decisions on what size boxes to use should be made, either 7-inch or 10-inch or a combination of both.

LEWIS: Documentation of archival treatment is a standard procedure in the conservation field. We should do no less in providing a paper trail of what was done and why, by whom, and what standards or equipment settings were used.

CORNETTE: Often the client wants a report from the engineer on the condition of a recording. For example, was the condition of the recording in poor, fair, good or excellent condition? If in poor or fair condition, how was the condition treated by the engineer in order to produce a preservation master? What were the technical processes used to make the preservation master such as baking reels to cure sticky-shed syndrome? What was the original date of the recording and what was the date of the re-recording?

The laboratory should work with the institution based on their current inventory control system, on how to keep a database and generate labels with the information required for their access purposes. The laboratory could create a database in MS Access for the institution's future use and then also work with the institution to develop the proper metadata for digital files for future access.

LEWIS: The goal of shipping is to insure the safety of the original materials as well as the copies made from them. NARA uses overnight shipping Monday through Thursday or in-person pick-up and delivery services.

CORNETTE: The laboratory should work with the client to determine which is the best way for their sound collection to travel. The laboratory can either hand carry or ship overnight the original masters of the sound collection. The original masters should always be shipped separately from the preservation masters and service copies. The shipments should be on separate days and the deliveries at separate times. The laboratory should carefully pack the sound collections or advise institutions how to pack the sound collections so that they are free from shock or vibration.

The laboratory should also be careful with magnetic items, so that they are not accidentally erased in travel. The laboratory should provide temperature controlled jolt-free transportation. Transportation is a large issue with sound preservation because many items are irreplaceable. Just as each collection is unique, so is the transportation to and from the vendor.

LEWIS: A vendor's laboratory may not have the same high security as (I hope) you have the collection in. What compromises are you willing to make and how will you inspect and monitor the facility to insure that the vendor is providing adequate physical and intellectual protection for original materials while in its custody?

CORNETTE: The laboratory should be as secure as possible with several locks and alarms. The alarm systems must detect for fire and for intruders. The building should always be locked at night and if the laboratory has some kind of guard or attendant during business hours for the building, that's always a plus. The laboratory might consider keypad locks that can block any access or reproduction to unauthorized people. Part of the hiring process should be conducting a background check of the laboratory's employees. A limited number of people should have access to the key codes.

Fire extinguishers (Class C or one suitable for electrical fires) must be in the laboratory ready to be used if ever necessary. The laboratory must ensure standard storage conditions for the sound collections while in the laboratory's facilities. A properly set up lab should be in a building that is well-constructed, well-located, and free of environmental hazards. The proper temperature and humidity to be maintained is at a consistent temperature of 60° F to 70° F with humidity values of 45% to 65%. The system should operate 24/7. Extreme changes in temperature and humidity greatly increase physical deterioration and can result in chemical changes and fungal growth in the materials of which the medium is composed. The laboratory should have a chart recorder to take daily measurements of both temperature and humidity.

The sound preservation laboratory should be located in an area free from where harmful vapors might be absorbed. The laboratory also should be free of food and the worktable should be clean of any foreign substances. The laboratory must be in a flood free environment.

The materials from the sound collection should be placed in their proper upright positions on strong shelves when stored in the laboratory. There should also be a fireproof cabinet in the sound preservation laboratory for valuable irreplaceable sound collections housed at the vendor's facility until preservation work is complete. It is good to have a fireproof safe.

LEWIS: Periodic reporting may not be required for short term or small projects but for large or long-term ones, you will need them to monitor progress and finances.

CORNETTE: What kinds of reports does the client want to see? How often? The client will probably require the vendor to send a temperature and humidity report weekly. Is there a receipt of materials report that the vendor needs to cross check when the materials are picked up from the client? Does the client want a quality control check report for every item in the batch or collection?

Modern communication with the client and vendor has made many preservation jobs go smoother. It is important to communicate by e-mail in order to have proper documentation of any changes to the SOW and project. E-mail has allowed the client and vendor to keep in touch during the course of a project. It is a quick and inexpensive way to communicate and it does not require a lot of one's time.

LEWIS: What will be the frequency of invoicing: Weekly? Monthly? As batches are completed? Some funds up front if equipment fabrication is needed? Some other scheme? Will there be penalties for late work? Will there be incentive payments for accelerated performance?

CORNETTE: From a practical, business point of view, it is better to bill in batches of a collection the laboratory is preserving, especially if the collection is large. Often times, this will coincide with a business calendar month for internal P & L reasons. If the laboratory waits until the entire collection is complete, it might be waiting several months or years. That's a long time to go without any revenue. The vendor should work with the client to decide what a reasonable amount of items to be preserved in a batch would be and then bill monthly no matter many batches were completed in the month. Clients often wait to pay the vendor when they have acknowledged receipt of the batch and new preservation masters or have completed quality control of the new preservation masters. If there is rework to complete, rework is completed before payment of the services is issued. This kind of invoicing allows the client and the vendor to keep track of money allotted to the contract and how many items are being done.

LEWIS: To wrap this up, we've talked about:

- ▶ Recognizing the three needs for lab services: failing recording media, obsolete technologies, and user demands.
- ▶ Some convenient operation definitions of conservation, preservation, and restoration.

- ► Some laboratory reproduction services.
- ► Creating a Statement of Work by which to solicit and manage a contract.

The Library of Congress Digital Audio Preservation Prototyping Project

Carl Fleischhauer
Project Coordinator
Office of Strategic Initiatives
Library of Congress

THE DIGITAL AUDIO PRESERVATION PROTOTYPING PROJECT was established at the Library of Congress for several reasons. The underlying motive—not always visible in our presentations—is that the time has come to change our approach to reformatting recorded sound collections, for reasons I will outline in a moment. The surface motive, the trigger to action, is the planned move in 2005 by the Library's Motion Picture, Broadcasting, and Recorded Sound Division to a new facility in Culpeper, Virginia. Substantial funding for the new National Audio-Visual Conservation Center comes from David Woodley Packard (the son of David Packard, co-founder of the Hewlett-Packard Corporation) and the Packard Humanities Institute.

The project has embraced sample collections from two Library of Congress divisions: the Motion Picture, Broadcasting, and Recorded Sound Division (M/B/RS) and the American Folklife Center (AFC). In another talk at the "Sound Savings" conference, the archivist Michael Taft described the AFC Save Our Sounds effort, which is allied with the prototyping effort. Overall, the project's focus has been on reformatting sound recordings, with an eye on moving into video. We want to reach some useful conclusions next year, in time to apply the lessons in the new building.

The prevalent practice for reformatting audio and video from the 1960s and 1970s into the 1990s has been "copy to analog magnetic tape." We see four reasons to change. First, there is the matter of media life expectancy. Magnetic tape (analog or digital) will not last as long as the archetypal media used for reformatting: microfilm. Second, there is the issue of quality loss as a result of making the copy. Analog-to-analog copying introduces what is called generation loss. This is tolerable with microfilm, when the time between re-reformatting is long. But with audiotape the time between re-reformatting is relatively short and the adverse effects are troubling. Third, there is the problem of device and media obsolescence. We are seeing a virtual cessation of manufacturing of analog-tape media and analog-tape recording devices. Finally, the digital era is here and we need to engage it, and not just to serve reformatting. The next generation of content to reach our institutions will be digital to begin with and its preservation for the long term will depend upon techniques similar or identical to those we establish to sustain digitally reformatted content.

The prototyping project has also been motivated by the desire to model new ways to provide access to researchers. The production of digital masters makes it relatively efficient to produce service copies, e.g., compressed copies that can be accessed in our secure local area network or streaming copies for the Web. At the Library of Congress, copyright considerations and consideration of the prerogatives of folk communities mean that we must limit access to much of our recorded sound collection, i.e., many items cannot be placed on the public Web. But after the collections have moved to Culpeper, we want our reformatted content to continue to be accessible in reading rooms on Capitol Hill, and the digital service copies that we place in the Library's secure storage systems will help us accomplish that goal. We are also exploring ways to provide access more widely, perhaps to remote sites, legal circumstances permitting.

One of the advantages of digital-file reformatting is the ability to reproduce an entire object. For example, here is a description of our digital reproduction of a sound recording made by the U.S. Marine Corps in the Pacific during World War II. The 1945 original was recorded on Amertape Recording Film, sprocketed 35mm film that ran through a recorder that cut grooves in the surface. Within a year or so, the Marine Corps copied the film to 16-inch transcription discs. These have since deteriorated but they were used as the source for our audio. (We hope to go back to the film at some point.) The digital copy provides access not only to the audio but also to images of the film box, the disc labels, and a content log sheet that had been packed with the film. This virtual package is presented in an interface that permits a researcher to play the audio, zoom in on the images, and examine detailed technical metadata.

The preservation approach we are exploring has at its core a *digital object* or *information package* that includes bitstreams, i.e., the files that contain the audio and images, and metadata. These packages will be managed in digital *repositories*, sophisticated versions of the computer storage systems we are using today. CDs or DVDs will not be used to store the content. It is worth saying that content management—what happens inside the repository—has at its heart a paradox. Digital content depends on specific information technology systems to keep it alive and to render it for users. But information technology systems are inherently obsolescent and will be replaced in relatively short time periods and thus our content must also be system independent. At any given moment, content lives on *this* media—disks in this server, for example—and is sustained by *this* information technology system, but the content must transcend the lifespan of any given media and system.

Our preservation explorations have wrestled with four issues: selecting the target format for reformatting, determining the quality of the reformatted copy, shaping the information package and the importance of metadata, and analyzing longevity in a "media-less" environment.

Selecting the Target Format

The first issue concerns the choice of bitstream structure and file type. This entails striking a balance between six factors:

▶ **Disclosure**: Are specifications and tools for validating technical integrity accessible to those creating and sustaining digital content? Preservation depends upon understanding how the information is represented as bits and bytes in digital files.

▶ **Adoption**: Is this format already used by the primary creators, disseminators, or users of information resources? If a format is widely adopted, it is less likely to become obsolete rapidly, and tools for migration and emulation are more likely to emerge.

▶ **Transparency**: Is the digital representation or encoding open to direct analysis with basic tools? Digital formats in which the underlying information is represented simply and directly will be easier to migrate to new formats. Encryption and compression inhibit transparency.

▶ **Self-documentation**: In part, this is about the package inside the package. Does the file format include metadata that explains how to render the data as usable information or understand its context? Self-documenting formats are likely to be easier to sustain over long periods and less vulnerable to catastrophe than ones that are separated from key metadata.

▶ **"Fidelity"** or support for high resolution: Does the format "hold" high resolution audio?

▶ **Sound field support**: Does the format represent stereo and even surround sound?

What formats have we selected? For our audio masters, our bitstream choice is pulse code modulated (PCM) sampling, uncompressed. This is the type of bitstream used on audio compact disks and it meets the transparency test. The file format we use is WAVE, from Microsoft, and it meets the adoption, disclosure, and fidelity tests. By the way, we feel that the "PCM-ness" of the bitstream is more important than the "WAVE-ness" of the file; Macintosh users put their PCM bitstreams into AIFF files to equal effect. We have not yet begun using what is called the Broadcast WAVE Format, which would get a higher score on the self-documentation test than ordinary WAVE. Meanwhile, we are curious about one-bit-deep formats like the DSD structure on SONY's Super Audio Compact Disk (SACD) but this bitstream structure is not yet widely adopted. DSD also gets occasional negative write-ups in the trade press, so we are taking a wait and see position. Since our reformatting is limited to mono and stereo material for the moment, we can afford to put off addressing the matter of surround sound. For audio service files, we use WAVE at lower resolution and MP3 compressed files.

For our image masters, our bitstream choice is bit-mapped or raster, also uncompressed. The file format we use is TIFF, another industry standard, originally from Aldus and now from Adobe. Here, the "bit-mapped-ness" of the bitstream is more important than the "TIFF-ness" of the file. For image service files, we use JPEGs and expect to switch to JPEG 2000 after this new format has been more widely adopted.

The Quality of the Reformatted Copy

The second central issue has been the subject of several interesting and instructive discussions by LC staff working on the prototyping project. Our talk revolved around questions like, "What does high resolution mean?" and "Why should we seek it?" In the end, our decision-making turned on some unexpected factors, some of which are beyond the reach of science and objective measurement.

With sound, the analysis of resolution starts with considerations of sampling frequency, measured in cycles or kilocycles per second. Roughly speaking, digitizing audio means taking the analog waveform and representing it as a large number of points or dots—connect the dots and you have your waveform back. The more dots, the better you can redraw your sound wave; the more dots, the better you can represent the fine parts of the curve that represent high frequency sounds. This parameter can be compared to spatial resolution for images. A digital image consists of row upon row of picture elements, pixels for short, often called "dots." The higher the number of pixels, the higher the spatial resolution.

The second key parameter is bit depth, which audio engineers sometimes call "word length." With audio, the more data you have for each sample—the longer the word, so to say—the more accurate the position of the sample in terms of amplitude. Greater bit depth gives you a lower noise floor and lets you represent a greater dynamic range, which can be especially helpful when transferring, say, field recordings made in hard-to-control circumstances. Compact disks usually have 16 bits (2 bytes) per sample, while many professional recording systems offer 24 bits (3 bytes). The imaging analogy is that an image 24 bits per pixel can reproduce more colors than 8- or 16-bit representations and thus offers the possibility of greater color fidelity.

Everyone is convinced that it is a good idea to digitize audio at 24 bits per sample. Keen ears can hear the difference and, although we have not done so, one could exploit test signals to compare distortion and noise. And it was in the discussion of bit depth that one of the "unmeasurable" factors was articulated: "You want a cushion of extra data," the engineers said, "just to protect you when you copy items with a wide or varying dynamic range, or to give you elbow room to fix things later in the event that an operator doesn't do a perfect job."

I have heard an analogous argument regarding imaging, especially when reformatting photographic negatives. The proposal is to make a preservation master image with a "flat" (low contrast) contrast curve and 12- or 16-bits-per-channel instead of the customary 8. Then a future user could manipulate the image to restore it or for a desired aesthetic effect, and resave it at 8 bits deep. The outcome of this process would be an image with a full set of tones at the 8-bit depth, i.e., the histogram for the new 8-bit image would be free of gaps. In contrast, if you started with an 8-bit-per-channel master, manipulated it, and then resaved the copy at 8 bits, the resulting copy image would lack some tones, i.e., the histogram would have gaps.

In contrast to the consensus we reached regarding the desirability of greater bit depth for sound recordings, our conversations about sampling frequency revealed differences of opinion. Some of us on the administrative side imagined that the starter question would be: "What is the range of sound frequencies that we might expect in this original item?" Our idea was that we would set the frequency range of the digital copy to more or less match the frequency range inherent in the original item. What frequencies have been captured, for example, on a 78 rpm disc from the acoustic era? From 8–10 kilocycles per second? The usual rule for digital sampling is to work at twice the highest frequency you want to reproduce. Therefore something on the order of 20 kilocycles per second should capture the full range of frequencies on the original 78. Or to take another example, suppose a collector used an analog Nagra tape recorder to record folk music at 7.5 ips with a Neuman condenser microphone. What is the highest frequency tone that we might expect to hear when the tape is played back? Most engineers would say that such a recording system is not likely to capture much sound with frequencies above 14 or 18 kilocycles per second. Thus if we digitally sample at 44 or 48 kilocycles, we ought to capture the full range of frequencies.

The engineers, however, did not want to work at 44 or 48 kilocycles, to say nothing of 20. They advocated 96, with some eyeing 192. The argument here—and really this argument takes both higher sampling frequency and greater bit depth into account—largely concerns factors that pertain to practical production matters or "downstream" possibilities, and which are therefore not very susceptible to objective testing. The following paraphrases capture some of the dialog:

- "There may be hard-to-hear harmonics that you won't want to lose."
- "Copies with less noise and less distortion can more successfully be restored when you come back to them."
- "In the future we'll have even better enhancement tools and post-processing, so save as much information as you can."
- "What if you need extra data to support certain types of resource discovery?"

The last bullet refers to what is called "low-level" features by the working group developing MPEG-7, an emerging metadata standard associated with moving image work. In the world of sound, low-level data would be used to support "find this melody" queries, for processes that produce transcripts of spoken word content, or to support a system executing the famous query "find me more like this one."

Thus the analysis of the inherent fidelity of the original did not provide the steering effect some of us had expected. Meanwhile, we also made test copies at higher and lower resolution and asked, "Can you hear the difference?" But these informal A-B tests also fell short of being conclusive. One engineer has proposed carrying out some empirical tests on post-processing actions to confirm the idea that the restoration of a recording (e.g., careful cleanup for publication as a CD) would be more successful if the master was at high rather than moderate resolution. But we have not yet carried out such an experiment.

The outcome is that our team generally works at the upper limit of available technology. We produce most of our audio masters at 96 kilocycles and 24-bit word length. At this time, we make two service copies: first, a down-sampled WAVE file at compact-disc specifications: 44.1 kilocycles and 16-bit words, and second, an MP3 file that is very handy in our local area network. Meanwhile, we produce images of accompanying matter, like disc labels, tape boxes, and documents. The master images are at 300 pixels per inch (ppi), with a tonal resolution of 24 bits per pixel.

Our project development has highlighted two additional topics that have to do with reproduction quality. The first has to do with practices, including the use of professional equipment and professional workers. On the equipment side, one key device is the analog-to-digital converter, the device that actually samples the analog waveform and spits out the bits. Professional converters are generally external to the computer workstation (or digital audio workstation) and are superior to and more costly than "pro-sumer" a-to-d devices, often installed as a card in the desktop computer. We avoid cleanup tools when making masters. And for mono discs in our collections, we copy using a stereo cartridge to allow for future processes to "find the best groove wall."

On the human side, digitizing requires professional skills in both the digital and analog realms. A professional worker must not only be conversant with a-to-d convertors and workstations, but must also be a master of the art and science of playing back originals to the best effect, no mean task when you confront instantaneous discs, cylinders, wires, and sticky tapes. In the new center at Culpeper, we see these professionals as our supervisors, contract overseers, and as experts who perform the most difficult work.

As we plan the future, we would like to include apprentice workers in the team, as well as outsource certain types of material. We have so many items in need of reformatting, that we are seeking ways to increase efficiency. Elements that we hope will accomplish this include sorting originals by "transfer efficiency" category, that is, by putting groups together that have the same technical characteristics. We would like to find and employ expert systems (automated tools) to help us judge quality or at least spot anomalies to inspect later. For some categories, we want to experiment with having a single operator copy two or three items at once. I will note that some interesting high-volume production tools are emerging from the PRESTO (Preservation Technology for European Broadcast Archives) project organized by broadcasters in Europe (http://presto.joanneum.ac.at/index.asp). At the same time, our team has been very interested to learn about Carl Haber and Vitaliy Fadeyev's cutting edge experiments at the Lawrence Berkeley National Laboratory to use high-resolution imaging to recover sound from discs and cylinders (http://www-cdf.lbl.gov/~av/).

The second additional topic pertaining to high resolution concerns the role of objective measurement. In imaging, this is related to the use of targets and, in audio, to standardized sets of tones. The outputs from targets or tone sets permit you to measure the performance of the equipment used to produce an image or an audio file and the setup or adjustment of that equipment. The measurement of targets and tones does not help you evaluate actual "content" images or sounds directly.

In library and archival reformatting circles, the development of imaging targets is farther along than practices for using audio tone sets. I participated in an image-related contracting activity at the library in 1995 and, at that time, the appropriate targets, the availability of measuring tools, and ideas about how to interpret the outcomes were not at all mature. Recently, experts have wrestled with what are called *performance measures* for digital imaging systems. You can't necessarily believe your scanner when it says 300 ppi, we are told. Instead, we should measure what actually comes through the system. For example, use modulation transfer function (MTF) as a yardstick for delivered spatial resolution. But the process of implementing performance measures for imaging has not yet reached its conclusion. My impression is that the investigators working on this are not ready to say what the MTF pass-fail points ought to be for, say, a system used to digitally reproduce a typical 8x10-inch negative.

On the audio side, our work group has made sound recordings of the standard ITU test sequences known as CCITT 0.33. There is one for mono and one for stereo, and both are 28-second-long series of tones developed to test satellite broadcast transmissions. With appropriate measuring equipment, recordings of the tones can be used to determine the frequency response, distortion, and signal-to-noise ratio produced in a given recording system. We have looked at the numbers but we are not yet ready to say where the pass-fail points ought to be for the equipment we might use. The recording industry may have more sophisticated or more appropriate performance measures, not well known in our circles, but I am sure that those of us working on the problem in the archive and library community will get smarter (or better informed) with time.

Shaping the Information Package and the Importance of Metadata

The third central issue concerns the *information package*, a complex multipart entity. As noted earlier, the package's data takes the form of audio, video, or image bitstreams, while its metadata represents a familiar trio from digital library planning: descriptive, administrative, and structural.

In our prototyping project, our main descriptive metadata is for the object as a whole and is often a copy of a MARC (MAchine Readable Cataloging) record in the Library of Congress central catalog. The copy is massaged to create a MODS XML record (http://www.loc.gov/standards/mods/). But our complex objects often benefit from additional descriptive metadata for individual parts, e.g., song titles, artists for disc sides or cuts, or names associated with a particular element within a digital package. The descriptive metadata for these elements are encoded in what MODS calls *related items*, a kind of "record within the record." One type is called a *constituent related item*, and this fits our case very nicely.

Our administrative metadata is extensive. For example, we include a persistent identifier and ownership information, meant here not in the copyright sense but rather to identify the party responsible for managing this digital object. We include information about the source item and any conservation treatment that may be applied, data about the processes used to create the digital copy (sometimes called *digital provenance data*), and technical details about the file we have created. In the latter two categories, we have made use of sets of data elements under discussion by working groups within the Audio Engineering Society; our versions of these data sets are linked from this Web page: http://lcweb.loc.gov/rr/mopic/avprot/metsmenu2.html. Meanwhile, we have not made a big effort to collect true rights data but we do categorize objects to permit access management at the library.

Structural metadata records the relationships between parts of objects. For example, when we reformat a long-playing record boxed set, we produce sound files for all of the disc sides, as well as images of the labels, the box, and the pages in the accompanying booklet. Thus our digital reproduction will include several dozen files and these are documented in the structural metadata. In the interface for end users, this metadata supports the presentation of the package and enables the user to navigate the various parts of the digital reproduction.

Although we have not implemented this in our prototyping, we know that there is a need for an additional category of metadata to support long-term preservation. This category is described in a helpful report from the Research Libraries Group and OCLC (Online Computer Library Center) titled *Preservation Metadata and OAIS Information Model* (http://www.oclc.research/pmwg/). Examples of digital preservation metadata include "fixity" information, e.g., checksums to monitor file changes; pointers to documentation for file formats; and pointers to documentation of the environment required to render files.

We are encoding all of the metadata using the emerging Metadata Encoding and Transmission Standard (METS) (http://www.loc.gov/standards/mets/). We worry about the extent of metadata that we wish to capture and count on the pressures of actual production to have a winnowing effect. Meanwhile, it is critical to continue the development of tools to automate the creation of metadata, especially administrative metadata.

Longevity in "Media-less" Environment

The fourth central issue highlights the importance of *keeping* digital copies, a need that rivals and may even surpass the need to *make* the copies in the first place. This is where the repository comes in, a topic of discussion rather than a point of action for our prototyping project. Regarding the repository, our project and the planning for the new National Audio-Visual Conservation Center intersect with library-wide digital planning—including a repository—being carried out by new National Digital Information Infrastructure Preservation Program (NDIIPP) (http://www.digitalpreservation.gov/). We anticipate that the design for the library's repository will be consistent with the important NASA Open Archival Information System (OAIS) reference model, now an ISO standard (http://ssdoo.gsfc.nasa.gov/nost/isoas/).

The OAIS model is the source of our *packaging* jargon. The model articulates a content life cycle in which a producer sends a *submission information package* to the repository, where it is ingested and reshaped to make an *archival information package*, suitable for long-term management. When an end user, called a consumer in the model, requests a version of the object for viewing or listening, the repository reshapes the content into a *dissemination information package* for presentation. We anticipate that the center at Culpeper will play the role of producer, preparing submission information packages for the library's repository.

During this period when the library's repository is under development, we place our carefully named files in UNIX file systems established in the library's storage area network. Although less sophisticated than the planned repository, the storage area network has an active backup system in place, a system that has sustained the eight million files from our American Memory program for several years. We keep planning improvements in our practices. For example, we now segregate our masters and service files so that a higher level of protection can be applied to the masters. For now, the METS metadata is stored as individual XML files. In effect, we are storing virtual information packages, "ready to submit."

There is a policy implication here. Keeping digital content requires a significant information technology infrastructure, meaning both systems and people. That may be fine for larger organizations but what about smaller or independent libraries and archives? Small sound archives are clearly not in a position to mount this level of IT infrastructure. What are they to do? There are two dimensions to this issue. Some future-oriented discussions in the NDIIPP context have suggested that there should be many libraries and archives—thought of as those who organize, catalog, and provide access to content—served by a *few* repositories—the keepers of the bits. This suggestion raised follow-up questions: How might such a many-few structure be established? Who would pay for what?

As these longer-term policy questions are being considered, there are pressing questions for today. Is there a suitable holding action for *keeping* digital content? For audio, would it be a good idea for small archives to store their files on multiple CD-Rs or DVD-Rs, or to write to data tape, as an interim solution? Ought one work in a hybrid manner, digital and analog, in spite of the extra cost? There are no authoritative answers for these difficult questions and this has impaired our ability to provide our colleagues with definitive answers.

Portions of this paper have been taken from a talk presented at the 2003 Preservation Conference at the National Archives and Records Administration (http://www.archives.gov/preservation/conferences/papers_2003/fleischauer.html). This paper represents work carried out for a federal government agency and is not protected by copyright.

Lyndon Baines Johnson Library and Museum Audio Reformatting Project

Jill Hawkins
Asheville, North Carolina

IN JANUARY 2002 I began a semester-long practicum at the LBJ Library and Museum working to reformat a collection of reel-to-reel audio tapes into digital files. I became interested in audio preservation during the course of my studies at the University of Texas at Austin's Conservation and Preservation Program. The practicum required research about current standards and reformatting procedures, the development of both standards and procedures to fit the needs of the reformatting project at the LBJ Library and Museum, and the implementation of these procedures by transferring some of the analog materials to digital files.

The LBJ Library and Museum was dedicated on May 22, 1971. It is part of a system of presidential libraries administered by the National Archives and Records Administrations. The library was established to preserve and make available for research the papers, photographs, audio and video materials, and memorabilia of LBJ's presidency. However, the library's holdings include materials from Johnson's entire public career, as well as materials related to his family and close associates. The library currently houses 40 million pages of historical documents. In addition, the library actively collects the papers of Johnson's contemporaries and conducts an oral history program.

The collection being reformatted includes the sound recordings of the president's speeches, press conferences, and public remarks known as the Presidential Speeches. There are 831 reel-to-reel tapes in the collection with a total of 2,100 speeches. The White House Communications Agency (WHAC) was responsible for most of the original recordings. When the speeches were original recorded, each one was put on its own individual reel. Each reel lasted approximately 30 minutes, so if the speech was longer than that, it was put on more than one reel. The length of speeches in this collection ranged from less than a minute to over an hour and a half. WHAC then combined different speeches of varying lengths to create reel-to-reel tapes with a combined length of approximately 30 minutes. The speeches were maintained in roughly chronological order based on how they fit together to form 30-minute reels. Thus 2,100 individually taped speeches were combined into an 831-tape collection known as the President's Speeches.

WHAC made many copies of the combined speeches including the master copy, the presidential copy, the working copy, the reference copy, and two archive copies. These copies are all currently stored on the third floor stacks at the library, an area that is kept cool and dry and monitored for both humidity and temperature. The lights in the stacks remain off unless someone is on the floor to retrieve an item. The original tapes are currently kept in the cold storage vault at the library and only used as a last resort. The president's copy is being used to make the transfers, with other copies being used only if necessary. A program of exercising the tapes every two years was begun in the early 1970s but this practice ended a decade later. The tapes have been stored tails out and overall are in very good condition.

The first assignment during my practicum was to create a digital finding aid to improve accessibility and establish better archival control. I designed and populated a Microsoft Access database that included the information located in the paper finding aid and then linked each event to the text location in *The Public Papers of Lyndon B. Johnson*. The database included additional fields to provide information, such as the file name and length, about the digital transferred files.

Another task during the practicum was to research options for outsourcing the reformatting project. While considering the specific needs of the library, I devised a questionnaire to follow when contacting a company about its reformatting services. Some of the library's specific needs and issues included: formats available for end product, quality of end product (file size and CD quality), experience with references, storage conditions, security issues, costs, and any additional services offered. One important factor was that the library wanted the work done locally. I reported my findings to my supervisor Philip Scott, Audiovisual Archivist at the LBJ Library and Museum, who presented the information to the Library Director, Betty Sue Flowers. The decision was made to perform the digital transfers in-house and I was offered the opportunity to do the work.

With the decision to do the reformatting in-house, I was asked to develop standards and procedures for the library. It was very difficult to decide what standards to develop because very little guidance is available and no "official" audio reformatting standards exist. There were several important factors I considered when developing standards for this specific project, including:

- Nature of material—spoken word
- Type of material—full track polyester reel-to-reel audio tape
- Condition and sound quality of the tapes—good
- Size of the collection—approximately 416 hours of materials.

The final standards and procedures were based on the needs and concerns of the library. We decided how to set appropriate sound levels for the event and then developed procedures to perform this task consistently. Standards to be developed included file size and file naming structures. We had to determine both the bit depth and sampling rate for the file size. We also had to decide how many generations of digital files to make and if preservation backup copies should be made on CDs. I researched the standards and procedures used by other archival institutions working on audio reformatting projects and looked for projects that were reformatting similar materials and institutions with national recognition. Conservation Online's Preservation of Audio Materials Web page[1] was a great source for information about audio preservation. The primary institutions I consulted included:

▶ The Library of Congress: Audio Visual Prototyping Briefing Document and Digital Conversion Statement of Work.[2]

▶ Michigan State University: Spoken Word Project. "Working Paper on Digitizing Audio for the Nation Gallery of the Spoken Work and the African Online Digital Library."[3]

▶ The Colorado Digitization Program, Digital Audio Working Group: Digital Audio Best Practices Version 1.2, May 2003.[4]

With any type of preservation, the goal is to preserve the material as closely as possible to the way it was originally created. Reformatting is a very invasive type of preservation. It results in material being transferred from one type of medium, in our case analog reel-to-reel audio tape, to another type of medium, digital WAVE files. This transfer inherently changes the material. The primary problem when determining what to do with audio materials is that currently an archival medium is not available for audio materials. However, for this project it was already determined to transfer the analog reel-to-reel tapes to digital files. A major benefit with digital files is that once the material is transferred, further sound quality will not be lost during additional digital transfers.

I had to develop a procedure for determining where to set the sound levels on the reel-to-reel machine and on the computer when performing the digital transfer. The procedure included the following three requirements: (1) the preservation copy should be a flat transfer of the original, (2) no digital changes would be made, and

(3) the sound level settings for the transfer would be the only element adjusted. For each event the sound levels would be set so that the loudest part of the event was just below clipping, and levels would never be adjusted during the event. Sound levels were checked by listening to the event and setting the levels on the reel-to-reel machine so that the VU meter levels were just below peaking. Then on the computer, the Line-In volume of the sound card was adjusted to make sure that the meters in Sound Forge also did not peak.

The next major decision was to set standards for the number of audio file generations to create and their file sizes. Some of the main issues and questions considered in making this decision included:

► How many generations are needed to insure the preservation of the audio materials? More generations cost more money because they demand more storage space on a server and more materials for CD backup copies.

► Digital capabilities are always improving so what is available now will relatively soon be out of date. For preservation purposes, the field generally tends to fall on the "high" end of the file size choices.

► Larger file sizes cost more money because they require more processing time and storage space.

► Is 24-bit justified over 16-bit because it helps to overcome imperfection in the digitization process?[5] Our tapes are in good condition and have relatively good sound quality.

► Is 48 kHz sampling rate justified over 44 kHz sampling rate because of the material type? For the same reason, is 96 kHz not justified for audio tape? The highest frequency pitch that a digital audio sample can hold is one-half of the sampling rate.[6] Thus, since analog audio tape is unable to capture frequencies above 24 kHz, a 48 kHz sampling rate has captured every thing possibly recorded on the audio tape. This reasoning seems to make 48 kHz preferred over 44 kHz and 96 kHz unnecessary.

We also had to weigh the benefits of higher file formats for our material, the spoken word. We performed an audio listening test to determine if differences could be heard between the different file size formats. The results were unclear; overall, differences were not readily noticeable. We devised standards based on the library's specific requirements and capabilities as well as our research of other institutions' standards. While we understood the importance of having a flat copy for preservation purposes, we were also hoping to address the issue of access. We decided to produce a reference copy so that, when necessary, we could adjust the sound level of the speech to make it easier for the listener to hear.

We decided to make three generations of files to be kept on the server:

1. **Preservation Master:** The file taken from the direct transfer of the reel-to-reel tape.
   - Uncompressed WAVE file format
   - 24-bit depth and 48 kHz (kilohertz) sampling rate
   - No enhancements

2. **Reduced Master:** Produced from the Preservation Master. In Sound Forge the file will be digitally reduced from 24 to 16 bit and the sampling rate will be reduced from 48 to 44 kilohertz.
   - Uncompressed WAVE file format
   - 16-bit depth and 44 kHz sampling rate
   - No enhancements

3. **Reference Service Duplication Master:** Produced from the Reduced Master. While listening to the original transfer, any noticeable sound level problems will be noted and then digitally corrected using the volume process in Sound Forge.
   - Uncompressed WAVE file format
   - 16-bit depth and 44 kHz sampling rate—same as reduced master
   - Volume adjusted when necessary to make the speech a uniform sound level
   - This generation will be used to make copies to sell to the public and eventually will be used to make MP3 files for Web use.

In addition to the three generations of files to be kept on the server, the decision was made to make three CD copies to serve as preservation backup copies:

1. **Preservation Master:** Data CD containing 24-bit and 48 kHz WAVE files. The WAVE file is uncompressed with no enhancements; it will serve as backup to the Preservation Master file on the server.

2. **Audio CD:** Made from the Reduced Master server files, it will be produced in "red book standard," also known as "CD standard." If nothing else survives, it is thought that one would at least be able to find a CD player that would recognize this copy.

3. **Reference Service Duplication Master:** Data CD 16-bit and 44 kHz WAVE files. This copy will have volume adjustments and will serve as backup to the Reference Service Duplication Master server file. The main reason for this backup copy is to preserve the work and time involved in performing the volume ad-

justment.

The last standard we wanted to develop was a file naming structure. While the exact file naming structure we developed is not necessarily important, it is crucial to set a standard for naming that is consistent over time. As with all other standards that are developed, naming structures need to be documented for future users. The specific formula used to develop the file name for the LBJ Library was as follows:

Example: WHCAAUDSP004t1pm.wav
  WHCA - White House Communication Agency, that created the material.
  AUD - Audio material type
  SP - Speech President collection
  004 - tape number (three numeric digits because tape numbers go into the hundreds)
  t1 - track number (collection contains tapes with 1 to 7 tracks per tape)
  pm/rm/rsd - file generation (preservation master, reduced master, reference service duplication master)
  .wav - file format

For the project, we used Gold Mitsui CD-Rs with a white inkjet printable surface for backup copies. The file name was printed on the CDs with black ink. The digital files are being stored on a server with access limited by password. The workstation set-up was put together by Fletcher Burton, Head of Technical Services. The specifics of the workstation are as follows:

  Reel-to-Reel Machine: Otari MX5050
  Pre amp: Peavey 10/4
  External sound card: Creative SB Extigy-Sound Blaster
  Dell Computer: Windows XP operating system, 1.4 MH processor and 1 gig HD
  CD printer: dedicated to only printing the CD surface (maker unknown)
  Jewel Case printer: HP 6122
  CD Burner: Que Fire
  Transfer software: Sound Forge 5.0
  Burning software: Nero-Burning ROM
  Printing software: Sure Thing CD Labeler used for both the jewel cases and the CDs.

Once there were general standards to follow and the equipment and software was set up, I began developing the actual transfer procedures. I started with a rough outline of each process in the procedure including (1) making the transfer, (2) burning the CD, and (3) making the jewel cases. Then I began to develop detailed step-by-step instructions on how to complete each process. The procedures were developed through a lot of trial and error. As I gained more familiarity with the equipment, the computer programs, and the collection, it became easier to determine the workflow. Three archival cabinets were purchased from Russ Bassett to store the three copies of the CDs that were being produced by the project. Two of the cabinets are located in different areas on the 3rd floor and the third cabinet is in the AV research reading room on the service level, one level below the first floor. When I left the project, I had completed 163 tapes including 454 individual speeches. One of my fellow colleagues, Sarah Cunningham, has taken over the project.

In the future, the library plans to sell the digital audio files that are being produced as part of this project. Currently, reference requests for the Presidential Speeches are being performed in the technical services department where they transfer the audio file from reel-to-reel to digital only on a demand basis. In the future, digital files created by this project will be used to fill reference requests. A CD will be made from the RSD (reference service duplication) server file into the format chosen by the patron. We are considering offering the following file formats: MP3, CD Audio, and a Data CD of the WAVE file. All requests will be produced from the RSD file, which has sound level adjustments, unless specifically requested otherwise. Currently, we are not making MP3 copies of the audio file because we are unable to have the sound files up on the Web; however, the library plans to make these sound files available. For the time being, MP3 files will be made on an as needed basis. The library also plans to make the digital finding aid available online. Staff would like to have space on the Web server designated for some of the audio files so that they will be accessible and downloadable online. Since this collection of speeches is in the public domain, the library does not have to worry about copyright issues. The Audio-Visual (AV) Archives division would like to purchase an AV Archives server that would act as a preservation server for all audio and photo files created by the division. This would allow AV Archives to gain better control of their preservation digital files. Currently the audio files have been allotted a certain amount of space, with password protection, on one of the storage servers for the library. The server is located inside the library's firewall. The library currently has several servers; one is outside of the firewall for Web-related information and the rest are inside the firewall and are used for daily activities and storage. Once this project is completed, there are many other audio collections within the library that still need to be transferred, including Lady Bird's speeches and LBJ's vice-presidential speeches. The LBJ audio reformatting project has provided a unique learning experience and hopefully will provide useful information for other institutions looking to reformat their audio collections.

Endnotes

1. Conservation Online, *Audio Preservation*. http://palimpsest.stanford.edu/bytopic/audio/.

2. Library of Congress, *Digital Audio-Visual Preservation Prototyping Project*, (23 March 2001). http://lcweb.loc.gov/rr/mopic/avprot/avprhome.html#coll.

3. Michigan State University, *Working Paper on Digitizing Audio for the National Gallery of the Spoken Word and the African Online Digital Library*. http://africandl.org/bestprac/audio/audio.html.

4. Colorado Digitization Program, Digital Audio Working Group, *Digital Audio Best Practices Version 1.2* (May 2003). http://www.cdpheritage.org/resource/audio/documents/CDPDABP_1-2.pdf.

5. Ibid.

6. Ibid.

Archiving the Arhoolie Foundation's Strachwitz Frontera Collection of Mexican and Mexican American Recordings

Tom Diamant
Digital Archiving Director
Chris Strachwitz Fontera Collection of Mexican and Mexican American Recordings
Arhoolie Foundation

CHRIS STRACHWITZ, A YOUNG IMMIGRANT from Germany, became enamored of the vernacular music he found when he came to the United States in the 1940s. He started collecting jazz, blues, and country 78 rpm recordings and by the late 1950s was also collecting Mexican and Mexican American recordings. The 78 collecting field had many fans of blues, jazz, and early popular music, but Mr. Strachwitz was one of the only collectors interested in Mexican music. He therefore ended up with the largest collection of Mexican and Mexican American commercial recordings. As he continued to seek out 78s, Mr. Strachwitz expanded his collecting to include 45 LPs, cassettes, CDs, videos, photos, and all things relating to Mexican music. In 1996 he donated the recordings of Mexican and Mexican American music to the Arhoolie Foundation. The archive continues to grow as Mr. Strachwitz finds more and more recordings.

The Arhoolie Foundation's Strachwitz Frontera Collection of Mexican and Mexican American Recordings, or the "Frontera Collection," as we call it, consists of an archive of almost fifteen thousand 78 rpm records, twenty thousand 45s, three thousand LPs, and numerous cassettes and CDs. In all, the Frontera archive totals over 110,000 individual recorded pieces. It contains many rare recordings including one-of-a-kind acetates and test pressings and many recordings of which we have the only known existing copy. This is a unique archive.

These recordings are all commercial recordings, but unlike many commercial recordings of jazz, blues, country, classical, popular, and other forms of "world" music, Mexican and Mexican American music has not been reissued much in the LP and CD era. Since the beginning of the recording era, commercial companies have been making Mexican recordings and selling them to audiences in both the United States and Mexico, but even though Mexican Americans now comprise the largest minority group in the United States, their recorded cultural heritage has been all but ignored by modern society.

Cultural Significance

The Frontera Collection offers a valuable window into the development of this culture as it has blended elements uniquely Mexican with styles, lyrics, vocabulary, social attitudes, forms, struggles, and politics that are clearly the result of the American experience. Within the Frontera Collection, this unique cultural heritage is preserved.

The Frontera Collection consists of *corridos, canciones, rancheras, boleros, sones, tangos, banda, polkas, light classical,* and *dialogues,* along with a whole range of American popular music interpreted by Mexican and Mexican Americans. The entire collection represents a window into the changes in American and Mexican American culture during the last one hundred years.

Many of these records never achieved significant distribution on a national scale, which in turn helped them maintain distinctive regional styles. This lack of assimilation led to the development of Tejano and Chicano music that continues to produce themes that are uniquely Mexican American. The lack of distribution and/ or wider American influence, however, also contributed to the lack of acceptance of this music as a valuable art form. Produced by scores of small regional labels, many of these recordings have virtually disappeared as the labels went out of business.

Steps to Preservation

The Frontera Collection is housed in a specially built building in the Arhoolie Records complex in El Cerrito, California. El Cerrito is located between two major earthquake faults, with many minor faults running between them. We are keenly aware that one serious earthquake or fire could destroy the archive. As soon as the Arhoolie Foundation took over the archive, preservation became a major concern.

The Archiving Process: Cataloging

The first step toward preservation was to catalog the archive. This was accomplished with grants from the National Academy of Recording Arts and Sciences, the National Endowment for the Arts, and Arhoolie Productions. A database was created and all the information from the record labels was entered. This information includes the record label, selection title, artist, catalog number, matrix number (the number given at the time of the recording by the record company), song type, and any other information that might be on the label or album sleeve. Now we knew exactly what we had in the collection.

One of the Arhoolie Foundation board members, Guillermo Hernadez, was not only concerned about the preservation aspect but also the accessibility issue. Over 110,000 recordings on the shelf in a building that only two people have the keys to is not a very accessible archive.

In 1999, Mr. Hernandez was the Director of the Chicano Studies Research Center at the University of California at Los Angeles (UCLA). He was interested in obtaining a copy of the Frontera Archive and depositing it in the Music Library at UCLA. This appealed to us at the Arhoolie Foundation. We would make two copies of the collection. One would be a protection copy to archive and make accessible at our facility and the other copy would be for UCLA to make accessible through their Digital Library system. Through their resources, UCLA could make the collection far more accessible to far more people than the Arhoolie Foundation, with our limited resources, could ever do.Through Mr. Hernandez's efforts and his association with the Noteño superstar band Los Tigres del Norte, the Chicano Studies Research Center received a grant and the promise of future grants from the Los Tigres del Norte Foundation with part of that grant specifically earmarked for the preservation and accessibility of the Frontera Collection. Los Tigres del Norte is a band that knows its musical roots and is very interested in making these recordings accessible to as many people as possible through UCLA. In 2001, the Arhoolie Foundation received a grant to start the project through the Los Tigres Del Norte Foundation, Chicano Studies Research Center, and the Fund For Folk Culture. This grant would finance the creation of a searchable, digital archive including a visual image of the label, a digital audio copy of the recording, and a discography, all accessible through the UCLA Digital Libraries Web site.

In the summer of 2001 we started our project. In collaboration with UCLA music librarians Gordon Theil and Stephen Davidson, we created a procedure to digitize the collection. We decided to start with the 78 rpm recordings.

The Archiving Process: Digitizing 78s

78s particularly are a wonderfully stable medium. They have a marvelous weight and feel to them and the grooves are so wide you can almost see the music in them. Unlike digital archival mediums, they can be played back on the most primitive of devices. You don't even need electricity. A scientist coming across one a thousand years from now could probably figure out how to get sound out of it without too much trouble. They are a very stable medium, until you drop or melt one. They are fragile in that respect.

We consulted with experts in the sound recording field, the archiving field, the digitizing field, and the data preservation field about digitizing techniques and procedures. We ended up with many varying opinions. Since this was and still is a new field with many different paths to follow, we realized that we needed to take the best advice we could, make our decisions based on our budget and needs, and just go for it. We knew that although we were a small foundation with a small budget, we could do an excellent and efficient job with careful, thoughtful planning and purchasing.

We decided to make a digital copy of the collection rather than copy it to analog tape. By digitizing it, we could create greater access to the archive, and after decades of experience in the recording industry working with tape, we quickly ruled out the unstable and obsolete medium of reel-to-reel tape. We would record directly onto the computer's hard drive.

Although those of us at the Arhoolie Foundation had great experience in recording 78 rpm records, we realized we need professional help when it came to digitizing them. With the help of Advanced Systems Group of Emeryville, California, a professional audio company, we put together a digitizing studio using the highest quality equipment we could afford.

Visual Documentation. We scan a high-resolution color image (at 300 dpi) of the 78 labels into the computer as well as make a recording of the 78. The scan includes the area around the label to the run-out groove. Often the matrix number, stamper number, or other information is inscribed in the record in this area. This allows the end user to not only hear the recording but also view a scan of the actual artifact and zoom in to read even the smallest of type.

With this image on the computer monitor, we compare the information with the information in our already existing database and correct any errors.

Preparing the Originals. We then wash the 78 with a specially built record-washing device. For some of the dirtier recordings, we use dish soap and run them under the faucet, taking care not to get the label too wet. The 78 is now ready to be digitized.

Digitizing. We digitize the recordings at the resolution 24 bits/ 96kHz directly into the computer and save them as uncompressed .wav files. The higher resolution was chosen because it simply captures a more complete representation of the audio waveform. It has also been suggested that if these digital files lose or have altered any of their digital bits as they follow their technological, migratory path, there is less chance for deterioration of the audio signal at the higher sampling rate.

To get sound into our computer, we needed the best analog to digital converter. This takes the analog sound and changes it into a digital signal that the computer can capture. Our biggest single expense was the external analog to digital converter. (See the Appendix for the equipment list.)

Slating. Upon recording each side of a 78, we created individual computer files of each recording in the .wav format. We decided to verbally slate each recording at the beginning by reading the label, catalog number, matrix number, title, and artist. This has proven to be both a great help and a bother. It is helpful since each individual file has the complete label information spoken at the beginning. Even without the discographical database, the recording is identifiable. When listening to a series of recordings in a row, it is also helpful to hear what each one is. However, if you are quickly going through the recordings using the interactive database, it becomes tiresome to hear the spoken slate. Fortunately, you learn how to skip over it.

Data Documentation. Since we started this project we have upgraded our software to use the Broadcast WAVE Format. This allows for all the information to be added to a text header in each individual sound file. We now not only list what information is on the label but the metadata of all the equipment used in the digitizing process right down to the stylus size.

Selecting Styli and EQ Curve. Our digital archivist tests the 78 with different diameters styli before choosing which one achieves the least noise and the best original sound out of the grooves. We use a pre-amp that can be adjusted to recreate a variety of equalization curves. There was no standard eq curve for mastering 78s. Play it back with the wrong eq curve, and it can muffle the sound or make it shrill and unnatural. Some eq curves are known for some of the major labels, but for the smaller, independent labels, and many of the major ones as well, careful, knowledgeable listening is the only way to know what setting to use.

The most important qualification for a digital archivist is a good set of ears. We hired a musician, Antonio Cuellar, not for his knowledge of formal archiving, but for his knowledge of what musical instruments and the human voice should sound like when they are played back through a good pair of speakers.

Keywords. As the recording is being played, the archivist listens and notes keywords from a list we have created. These keywords tell us what the main idea of a song is and which instruments are featured in an instrumental. We have keywords such as labor, migrant labor, revolution, murder, contraband, intoxication, and saxophone. Since most of the songs are love songs, we have many qualifiying words for love, such as praise, betrayal, jealousy, and murder. These keywords will be used by end users to search for recordings under a variety of topics.

Notation Field. A notes field is used to log additional information. Any anomaly of the record is noted, such as the labels on the wrong side, or incorrectly labeled all together, whether it is chipped or cracked, or excessively worn or too damaged to be recorded. If the recording is of a corrido, the Mexican narrative ballad form, further details are noted about the song, particularly if the song is about a true event. Due to limited time and the expectations of our funder, we cannot be as detailed as we would like in our notes field. Ideally, we could have more details about each song.

Results. We end up with a label scan, digitized sound file, and a database with discographical information and key words. How do we archive these at the Arhoolie Foundation and how do we send them down to UCLA to add to their digital library?

Data Archiving and Data Transfer Method. At the suggestion of our sound consultants, Advanced Systems Group, we chose to burn the files to DVD-R. For transferring them to UCLA, DVD-Rs are a very stable medium. Unlike a portable hard drive, you can drop DVD-Rs and there is no damage. They are not affected by magnetism and they hold 4.7 gigabytes. They are a standard storage medium that any modern computer DVD drive can read on any platform. UCLA then copies the information from the DVD-R onto its Digital Library mainframe computer.

Storage at The Arhoolie Foundation. For the Arhoolie Foundation Archive, we wanted a medium that entails the least trouble and expense to store the 30,000 individual recordings that comprised the 78 collection. Hard drives were too expensive at the time and required too much maintenance. CD-Rs did not store enough per disc. We decided on DVD-R, which holds seven times more information than a CD-R.

The question was how long would a DVD-R last? If you listen to the manufacturers, they will tell you up to 100 years. I can tell you that the ones we burned three years ago still work. Selecting the medium to use for storage was one of the most difficult to determine. What is commonly assumed is that any files we create will have to migrate to the newest and latest storage device periodically to keep the storage medium from becoming obsolete. For now we burn a DVD-R and put the copy on the shelf. At some time down the line, we will probably have to copy them all to a new medium.

We have the luxury of knowing that UCLA will be putting the files on their Digital Library Computer System, backed up regularly by the UCLA IT staff. Even if there were a disaster at the Arhoolie Foundation, there would be the copy at UCLA and vice versa. This is one of the key advantages of digitizing an archive. Multiple copies can be easily made and stored in multiple locations guaranteeing that at least a digital form of the archive will always be preserved.

User Access. The UCLA Library has developed a searchable, interactive Web site http://digital.library.ucla.edu/frontera/ where a user, logging in from the UCLA internet domain, can have complete access to the discography, sound recordings, and label scans. This allows an end user to search on a variety of criteria, read the discographical information, listen to the recording, and view the label. Due to copyright laws, a user logging in from outside UCLA can listen to only the first sixty seconds of the recording but can still view the label and discographical information.

UCLA is continuing to explore ways to make the archive more accessible. Researchers, students, and ordinary fans of the music, both at UCLA and around the world, are already using this Web site. By the time we are finished digitizing the 78s in the spring of 2005, UCLA will have almost 30,000 individual recordings accessible from this Web site.

In-house Database at The Arhoolie Foundation. The Arhoolie Foundation has created an accessible database for in-house use. We have converted all of the sound files into MP3s so that the file size is small enough to fit on a large single hard drive. We have also made medium resolution JPGs of the label scans. We have designed a searchable database. The database takes full use of the information in each data record, the label scans, and the sound files. When you find a record in the database, there are buttons to click on to view the label and hear the recording. You can search on any of the fields: artist, keyword, composer, musical genre, label, etc., and sort by these as well.

This makes studying a collection—once only accessible by going to the shelf and listening to one 78 at a time—far more easy. If you are looking for corridos about migrant labor, performed by Los Madrugadores on the Vocalion label, you can find it, listen to it and view the label as fast as you can type the query. Or you can quickly browse and listen to any recordings found by using any criteria.

One of the most interesting benefits of digitizing our collection is that we are actually listening to every 78 in the archive. This collection, like many other archives, was often added to by obtaining large quantities of 78s in bulk. Until now, no one has ever listened to each individual recording. Through the process of digitizing and noting keywords, we are carefully listening to each and every recording in the collection. The result of this is now we really know exactly what is in the archive and we are discovering many hidden gems. By having, for the most part, one person, Antonio Cuellar, listening to all of these recordings, we are also creating quite an expert in the field on Mexican and Mexican American Recordings.

The Future: A Digital Enyclopedia

As we digitized the Frontera Collection, we realized that this was just the beginning of a digital encyclopedia on Mexican and Mexican American recordings. We have in our collections hundreds of photos of musicians that are being scanned and added to the archive. We have artists' biographies, record label histories, catalogs, recording logs, lyrics, articles about the music, posters, and videos. We are working on digitizing all this information and linking it together so that as you listen to a recording, you can view photos of the artist, read a biography of the artist, a history of the record label, view a poster of a concert, look up the recording log and see how much the musicians were paid for the recording session, read the lyrics, and even view a video or listen to an interview of the recording artist, getting an entire view of the music and where it came from. This way of examining an archive gives the end user as complete an educational and emotional experience as a computerized experience can be. It may not be the same as holding the actual artifact in your hand and listening to it on a turntable, but in some ways it can be much richer.

Appendix

Equipment used for the Arhoolie Foundation's Strachwitz Collection of Mexican and Mexican American Records Digital Archiving Project.

Technics SP-15 Pro-Base Turntable
Shure M44G stereo cartridge
Styli in the sizes: 2.0, 2.3, 2.5, 2.8, 3.0
Optimus Sound Mixer SSM-1250 & Shure SM-57 Microphone
    (For slating recordings)
Elberg MD12 – Multicurve Disc Phonograph Pre-amp
Prism Sound Dream AD-2 Analog to Digital Converter
Rotel Stereo Integrated Amplifier RA-972
Miller and Kreisel MPS-1610 Nearfield Monitors
Keith Monks Record Cleaning Machine
Epson Expression Scanner 1640 XL (12 x 17 scanning bed)
Hp Desk 960c Ink Jet Printer
Maxtor External 120 gig Hard Drive
    (For in-house accessible database and MP3s, JPGs, photo archive, etc.)

Digital Recording Computer:
    White box Pentium 3–1000 mHz computer
    20 gig hard drive
    Digital Audio Deluxe Sound Card (This is used only to get the digital signal into the computer. We use the direct digital inputs since the analog to digital conversion has already taken place in the outboard Prism Sound Dream. We do not use the Audio Deluxe A/D converter.)
    Trinitron Sony Flat Screen Monitor

Scanning and DVD Burning Computer:
    White box Pentium 3–1000 mHz computer, 20 gig hard drive
    Pioneer A03 DVD-RW
    Trinitron Sony Flat Screen Monitor

Software:
    Wavelab 4.0–Sound Recording and Editing Software
    Filemaker Pro 6.0–Database Software
    Silverfast AI–Scanning software
    Photoshop 6.0–Image editing software
    Prassi Primo DVD burning software

Copyright Law and Audio Preservation

Georgia K. Harper
Manager, Intellectual Property Section
Office of General Counsel
University of Texas

STRICTLY INTERPRETED, the task of "preserving audio collections" encounters few if any copyright problems. After all, the Copyright Act specifically authorizes libraries and archives to preserve materials in any medium, including sound recordings, both published and unpublished. But preservation alone, without some context, may seem pointless. We don't wish just to preserve, but to provide access—at a minimum, to researchers—but more broadly, to the public, to the people whose culture our audio collections record.

The Copyright Act (17 USC 101 et. seq.) defines what will be protected and for how long, the rights of copyright owners, and limitations on those rights in favor of the public. Copyright owners have a set of exclusive rights that allow them to control how their works will be exploited—to a point. Their rights are not absolute, but are limited by the rights the act provides to members of the public to use the works, and to special groups such as libraries and archives, that make works available to the public, and by a term of protection that does not last forever (it only seems that way). If there is no specific authorization in the law for a use one wishes to make of another's work, that use will require the permission of the copyright owner. Once works enter the public domain, they may be freely used by anyone for any purpose. The date on which works enter the public domain varies depending on when and where they were created, but as a general matter, the term for protection averages and in many cases exceeds 100 years in the United States.

The framework of copyright law treats different kinds of rights, different kinds of media, and different kinds of uses differently. Every little subtlety is important and is there for a reason. Sometimes, however, the reasons get lost over time so that things that made sense once no longer do.

The rather convoluted way sound recordings have been protected in the Copyright Act illustrates these points. They represented a "new technology" at one time and were initially denied protection altogether. When they were federally protected in 1972, their protection was woven into a pre-existing fabric that had evolved mainly to serve the needs of commercial book and sheet music publishers. Congress's focus was primarily on preventing piracy, that is, unauthorized duplication and distribution, so it did not give the owners of copyrights in recordings a performance right. Anyone could perform any sound recording without permission from the owners of the recording, though they were required to pay a statutory license fee to the owners of copyright in the underlying composition. The Internet prompted a change in this limitation, however, and recordings now enjoy a limited performance right by means of digital audio transmissions. The way recording copyrights have varied over time gives their protection a sort of "patch-work" quality that makes understanding difficult, even for lawyers.

For the next 20 minutes, we'll look a little more closely at the patchwork of protection for recordings as a way to help us figure out which recordings may be in the public domain. Then we'll focus on the special exemptions that permit us to archive and make recordings available to the public even when they are still protected, and finally, we'll assess the shortcomings in all of this and review some actions we might take to maximize what we and our patrons can do with the recordings we have so carefully preserved in our collections.

The Legal Framework

U.S. Recordings Protected by Federal or State Law. Today, federal law protects U.S. sound recordings fixed on or after February 15, 1972 (hereafter, "1972"). As indicated above, these recordings enjoy only a limited set of rights. Copyright owners have the exclusive right to reproduce a recording in phonorecords, to prepare derivatives, and to publicly distribute phonorecords. Recording copyright owners do not enjoy an exclusive right to publicly perform their recordings; rather, Section 106(6) provides that the owner of rights in a sound recording has the exclusive right to perform the recording publicly by means of a digital audio transmission. Over-the-air broadcasts by terrestrial broadcasters licensed by the FCC do not violate the limited right because there is an express exception for them. Further, there is a statutory license available for transmissions that are not interactive such as typical Web casting. Congress equates interactive transmissions with sales of records and treats them differently from "streaming" and Web casting that are primarily intended for listening. Copyright owners may authorize or forbid interactive transmissions, as they see fit, just as they authorize or forbid duplication and distribution of physical recordings.

So, what about all those pre-1972 recordings? It is tempting to conclude that U.S. sound recordings fixed before 1972 must be in the public domain since they are not protected by federal law, but that's not quite true. Pre-1972 recordings are protected by state statutes and common law copyright, misappropriation, and unfair competition laws until 2067 (95 years after the date on which recordings first received federal protection in 1972). I researched Texas's common law of copyright. What little law we had addressed the right of first publication and was superceded by federal law in 1978, the year we moved from the old dual system of common law copyright for unpublished works and federal statutory copyright for published works to a unitary system where federal law protects all works from the moment of their fixation in a tangible medium. Some states have more elaborate statutory and common law. New York, home of the publishing industry, has many cases that discuss the subject. While it is beyond the scope of this short overview to delve into the laws of all 50 states, we can note that for the most part, state laws today protect the owners of properties not protected by federal law from misappropriation and unfair competition. The hallmarks of a claim under state law are bad faith, fraud, or misappropriation coupled with copying and distributing someone else's property. Usually, a directly competitive relationship between the owner and the user is required. Mere duplication alone is not likely actionable.

Foreign Recordings Protected by Federal Law and Possibly State Law. Foreign recordings are protected in a different manner. Until 1996, pre-1972 recordings created and published in foreign countries were ineligible for protection in the U.S., just like their U.S. counterparts. Today, a foreign sound recording *not in the public domain in the country of its origin on January 1, 1996*, when the Uruguay Round Agreements Act (URAA) of the General Agreement on Tariffs and Trade (GATT) went into effect, *and*

- first published before 1972,
- in an eligible foreign country (one of the signatories to the Berne Convention or WIPO Copyright Treaty, a member of the WTO, or an adherent to the WIPO Performances and Phonograms Treaty),
- with at least one author or rights holder being a national of or domiciled in an eligible country,
- and not published here within 30 days of the foreign publication,

is protected in the U.S. for the full term of protection it would have had if published here as a book or image or other work comprising protectable subject matter under federal law—95 years from the date of publication. The URAA "restored" the foreign work's copyright in the United States.

Here are two examples that illustrate how this works:

▶ If a composition were created in 1920 and performed and recorded in London in 1935, the recording would be protected by U.K. law (Copyright Act of 1956) for 50 years from the end of the year in which it was first published, that is, until the end of 1985. So, on January 1, 1996, when URAA went into effect in the U.S., that recording was in the public domain in the U.K. and would not be eligible for restoration in the U.S.

- ▶ Another composition created in 1920 and recorded in London in 1947 *would* qualify for restoration because the recording would still have been under U.K. protection on January 1, 1996 when URAA went into effect. It would be protected for 95 years from its date of first publication, or until the end of 2042, even though in the U.K. it would have become public domain in 1997.

A pre-1972 foreign work that does not qualify for restoration might possibly be protected by state common law in the U.S., but courts should be reluctant to use state law to protect a work whose copyright has expired. See, for example, *Capitol Records, Inc. v. Naxos of America, Inc.*, 2003 WL 21032009 (S.D.N.Y.) and a recent Supreme Court case in which the Court came down strongly against the use of trademark law to protect a copyrighted work in the public domain, *Dastar Corp. v. Twentieth Century Fox Film Corp.*, 123 S.Ct. 2041 (June 2, 2003).

The URAA was challenged as unconstitutional in *Golan v. Ashcroft*; the case was stayed pending a decision in another case that dealt with an overlapping issue (*Eldred v. Ashcroft*, 123 S.Ct. 769 (2003)). *Eldred* was decided January 15, 2003; however, there has been no activity on *Golan* since that date.

Composition Copyrights. A further complication results from the fact that the composition embodied on a recording is itself protected by copyright, so in some cases, even if a recording is in the public domain, the underlying composition may not be, and vice versa.

The Public Domain

So, what may we conclude is in the public domain? Compositions published in 1922 or earlier and performed and recorded (or, in some cases, released) in foreign countries on a date that would result in their being in the public domain on January 1, 1996, in the source country (i.e., during or earlier than 1945 for countries that protect sound recordings for 50 years) are in the public domain, and we should be able to do with them as we wish and permit our patrons to do the same. There may be some question about possible state law claims for unfair competition, but as the *Naxos* and *Dastar* cases suggest, other laws should not be stretched to make the good faith use of public domain materials actionable.

U.S. recordings of public domain compositions (i.e., those published during or before 1922) fixed in a tangible medium before 1972 are protected by state laws, but as noted, state law causes of action center on unfair competition for the most part. That is unlikely to affect archival and research activities. Patrons who would like to commercially exploit such recordings, however, would be well advised to obtain permission from the owners.

There is one other category of works that may be in the public domain: works whose copyrights have been abandoned or waived. Before 1989, when copyright notices were required to claim federal protection, failure to affix the proper notice for publication resulted in the works becoming public domain. U.S. recordings published between 1972 and 1989 may be in the public domain if they were published or otherwise widely distributed without the proper notice. Also, works published between 1923 and 1964 had to be renewed at the end of their 28-year terms. If the copyright owners failed to renew the copyrights, those works would now be in the public domain. By some estimates, as many as 95% of copyright owners did not renew their copyrights.

Our Archival Rights Under the Copyright Act

Section 108. Libraries have rights to archive and distribute works that are protected by the law, but when we wish to make and distribute digital archival copies, the rights do not extend much beyond the walls of the physical structures that hold our collections.

Sections 108 (b) and (c) of the Copyright Act provide our basic authority to archive and distribute works. Section 108 (b) applies to unpublished works and section 108 (c) applies to published works. The main differences between the two sections relate to the purposes for which we may make copies and the circumstances that must exist to justify making them.

We may copy unpublished works for preservation and security or for deposit for research use in another public or research library if we currently possess the work in our library. We may copy published works only to replace them if they are damaged, deteriorating, lost, or stolen, or if their format has become obsolete, and only after we have determined after reasonable effort that we cannot obtain an unused replacement at a fair price. One crippling condition applies to both types of works: digital copies cannot be made available to the public outside the premises of the library.

Section 108 (h) provides libraries and archives with special rights during the last 20 years of a work's copyright term. If a published work is not being commercially exploited, a copy cannot be obtained at a reasonable price, or the copyright owner has not notified the Copyright Office that the work is being commercially exploited or copies are available at reasonable prices, the library may make digital copies for preservation, perform them publicly, and distribute them to the public for scholarship or research.

The language of 108 (h) is very generous, giving libraries rights to copy, perform, display, and distribute publicly even digital copies of published works in their last 20 years of protection. However, section 108 (i) says that the rights of reproduction and distribution contained in 108 do not apply to musical works except for preservation. This creates a conflict: 108 (h) says we can copy, distribute, display, and perform published works publicly, and music is clearly "performed," but 108 (i) says that 108 (h) does not apply to music (by failing to include 108 (h) in its exceptions). Frankly, I think this is just sloppy drafting. Even assuming it does apply, section 108 only covers published compositions in their last 20 years of protection and U.S. recordings published before 1972 that are not protected by federal copyright law. So, libraries can (arguably) archive digitally and distribute publicly for scholarship and research pre-1972 U.S. recordings of compositions that were published between 1923 and 1927, inclusive. Next year, the eligible dates will be from 1923 through 1928, and so on, each year adding another year's worth of works to the list.

Notice how the beginning date of 1923 stays the same; this results from the addition in 1998 of an additional 20 years of protection to all works then under copyright (the Sonny Bono Copyright Term Extension Act). Works published in 1923 that would have gone into the public domain in 1998 did not lose their protection, and for the following 19 years, no works will enter the public domain in the U.S. Thus, the next time a U.S. work will enter the public domain will be at the end of the year 2018, when protection for the works published in 1923 will finally expire—unless, of course, Congress passes another extension, which the Supreme Court, in *Eldred v. Ashcroft* cited above, has indicated it is free to do.

The only foreign restored works to which this "last 20 year" provision would apply would be those from countries, if any, whose laws protected recordings for 68 years or longer in 1996. A recording that was published in 1928 in a country with a 68-year term would still have been protected in 1996 in the source country and would have then been eligible for restoration. The restored term of protection in the U.S. would end in $(28 + 95 =)$ 2023. Such a work would be in its last 20 years starting in 2003. Next year, works published in 1929 in that country would enter their last 20 years of protection.

I am not sure that there are countries that protect recordings for 68 years or longer, however, because Article 18 of the Berne Convention requires that member countries provide a 50-year term of protection to pre-existing works originating in another WTO member country if those works have not already enjoyed a full term of protection in both countries. So the norm is 50-year terms, not 68- or 70-year terms. A work from such a 50-year term country would have to have been published no earlier than 1946 to be still under protection on January 1, 1996 when URAA went into effect. A work published in 1946 would enjoy a 95-year term (46 + 95 = 2041) in the U.S. Its last 20 years of protection will not begin until 2021.

Because recordings embody at least two copyrights, the composition copyright and the recording copyright, those protected by U.S. federal law (foreign restored works and U.S. recordings published after 1972) would arguably require that both copyrights be in their last 20 years to qualify for digital distribution pursuant to section 108 (h). This point has never been raised, to my knowledge, so there is no legal guidance.

Fair Use and Other Exemptions. In addition to libraries' rights to archive works, libraries enjoy fair use rights as well. Section 108 specifically states that nothing in that section affects the rights of fair use, so it is reasonable to assume that fair uses may be made even of digital archival materials that would otherwise be restricted to the premises. For example, digitized archival materials may be put on electronic reserves, incorporated into class projects by faculty and students, and performed in the classroom and in distance education in accordance with fair use.

Promoting our Patrons' Uses

Generally. Although the law gives us reasonable rights to create archives, what we can do with them, and more particularly, what our patrons can do with them, is tightly constrained by copyright. In particular, the restriction to building-only use for digital archival copies of analog works tremendously limits the research value of digital archives, to say nothing of their value to the public. We are fast approaching a time when for many people, what you can't find on Google does not exist.

This unhappy circumstance suggests first that we should proactively manage our collections and copyrights to facilitate patrons' uses. Even though we may not be authorized to distribute a work digitally, we can make use of public Web sites as well as the more typical proprietary indexing and finding aids to make our hold-ings known to the public. Without being overly aggressive, we certainly can try to negotiate the widest scope of rights in materials we acquire from a rights holder, most importantly, the right to provide digital access to researchers at a minimum, and ideally, to the public. We can update our acquisition forms to encourage broad grants of rights to access and use. We should revisit collection restrictions periodi-cally to encourage that they be lifted. We should not only obtain, but try to main-tain accurate contact information for copyright owners when we acquire materials without rights so that we can facilitate the permissions process; alternatively, we might maintain specialized resources to help patrons locate and contact copyright owners. We may even try to acquire more general rights from a rights holder in connection with a specific patron request. We might create materials that can be used to educate researchers about the issues of rights when they are involved in the creation of recordings and forms they can use to clear rights expeditiously. Anthony Seeger speaks more extensively to these issues in "Rights Management—Intellectual Property and Audiovisual Archives and Collections" at http://www.clir.org/pubs/reports/pub96/rights.html.

Lobbying for Changes in the Law to Promote Access. Ultimately, we must bring the issue of greater access to digital archives to Congress. The changes made in section 108 in 1998 were intended to address serious problems—the complete lack of a digital archival right and the effect on research and scholarship of a new 20-year extension of the copyright term—but they addressed these problems in a way that only opened the door a crack. Copyright owners' fears that wide access to digital copies in libraries would undermine sales of the owners' works prevented adoption of any but the most restrictive provisions. But today, the National Recording Preservation Board has a mandate from Congress to study laws that must be changed to make preserved recordings available digitally, 2 USC 1724 (b)(4). Sections 108 (b) and (c) must be at the top of their list. If they are not, the board needs to hear from librarians and archivists about the need for changes in this law.

Summary

In response to a request from Ann Blonston, via e-mail before the conference, I offer as a summary this checklist of procedures to assess the extent of the rights you have to digitize the audio recordings in your collections and make them available to researchers or the public:

1. Determine whether the **recording and underlying composition** are in the **public domain** or protected only by state unfair competition and similar laws.
   a. Composition published before 1923; recording created and published in a foreign country on a date that would result in its being in the public domain in the foreign country in 1996. For countries with a 50-year term of protection for recordings, the date would be during or before 1945.
   b. Composition published before 1923; recording created or published in the U.S. before 1972. **Note:** archival uses are very unlikely to be actionable under state law; patron commercial uses probably need permission from recording copyright holder.
   c. Compositions published between 1923 and 1964 in the U.S. whose copyrights were not renewed and compositions published between 1923 and 1989

whose copyrights were abandoned by failure to follow formalities; recordings created and published in the U.S. before 1972 (see note above at 1b).

2. If the composition is still protected in the U.S., **digitize and archive recordings in accordance with sections 108 (b) or (c)** depending on whether the composition at issue is unpublished or published, respectively. Digital copies of analog works cannot be made available to the public outside the premises of the library or archive.

3. If the composition is in its **last 20 years of protection** (i.e., published between 1923 and 1927 inclusive), and the recording is pre-1972, digitize and archive U.S. recordings in accordance with § 108 (h) (see note above at 1b).
   a. Determine that the work is not enjoying normal commercial exploitation.
   b. Distribute it for research and scholarship with no limitation to the premises.
   c. For commercial uses see note above at 1b.

4. For proposed uses that exceed the limits imposed by section 108, consider whether **fair use** or another exception (TEACH Act, for example) may apply.

5. If no exception covers a proposed use of a protected work, it requires the **permission** of the owner of copyright in the recording and its underlying composition.

6. If you find that many uses you or your patrons would like to make of your collection fall outside the scope of statutory authorizations, consider more **aggressive rights management** strategies to secure broader rights for public use when possible **and legislative initiatives** to broaden the scope of current authorizations.

Shaping an Education and Research Agenda (Panel Session)

Mark Roosa
Director of Preservation
Library of Congress

THE FINAL SESSION of the symposium focused on shaping an applied research and education agenda. To address this area of concern, a panel of experts, composed of Tom Clareson, OCLC; Carl Fleischhauer and Sam Brylawski, Library of Congress; Ellen Cunningham-Kruppa, The University of Texas at Austin; Alan Lewis, National Archives and Records Administration; and Sarah Stauderman, Smithsonian Institution, discussed a series of questions posed by moderator Mark Roosa. The questions were:

1. What are the two to three most pressing preservation problems we face (in the field of audio) that could be informed by research and education?

2. What sorts of research projects (technical studies, demonstration projects, etc.) might be configured to address these problems?

3. Where might such research take place?

4. Who might be the leaders and collaborators?

5. How would this benefit the field and how would results be shared?

6. Who might be interested in funding this work?

Panelists shared their views with the audience and answered questions. A rich discussion followed which produced a substantial number of ideas concerning the current state of research and education and some valuable thoughts on what might be done to address these concerns.

Education and Training

There was an enthusiastic consensus among panelists on the need for a core educational curriculum for audio preservation to be used in library schools and in conservation training programs. Ellen Cunningham-Kruppa (The University of Texas at Austin) suggested that such a curriculum should build on the well-established and proven preservation approaches and philosophies that are taught in conservation and preservation graduate training programs around the world today, and not be developed independently of these approaches, which have proved successful in training professionals to meet the changing preservation challenges affecting library and archival collections.

At a minimum, the core elements of such a curriculum should include courses in:

- History, philosophy, ethics of preservation
- Science and physical characteristics of materials
- Storage and handling of materials
- Remedial preservation (e.g., reformatting, physical treatment, etc.)
- Management and administration of programs and projects
- Selection for preservation

Carl Fleischhauer (Library of Congress) added that a well-rounded curriculum should also include technically-oriented courses on topics such as audio transfer, understanding the audio signal chain, and courses that build computer literacy, such as how to set up a database, how to command UNIX, and how to use XML.

Mr. Fleischhauer emphasized the growing need within the field for individuals with technical skills in preparing audio materials for digitization. He also cited the need to train students in, for example, XML markup language. Another area of critical importance is developing skills, expertise, and tools for signal extraction from analog carriers and reliable methods for measuring the quality of a digital transfer. Recognizing that extracting signals from analog carriers is "a little bit art and a little bit science," one panelist suggested the need for individuals doing this work to be able to answer the question, "How do I know when I have all of the signal?" More training is needed, not only in selecting which tools are generally best suited for extracting signals from sound carriers (styli, pieces of equipment in the transfer chain, etc.), but also in the area of critical listening and perception. Several participants suggested that internships and fellowships for individuals interested in developing practical skills in audio preservation would be a way to begin to address this need.

Concern was also raised about the growing number of technical experts with extensive knowledge of the older antique formats and playback systems who are retiring (or about to retire) and will take with them intimate knowledge about how best to use and maintain these older analog systems. One way of transmitting this knowledge to the next generation of practitioners might be to hold a series of "summer camps" for audio technicians, where a small group of individuals could work directly with a senior engineer in a laboratory setting on a variety of technical issues (stylus selection, tape machine maintenance tips, etc.). It was suggested that several locations with audio labs (i.e., Library of Congress, University of Texas, Indiana University, etc.) might initially host such camps. As an adjunct to this approach, "catch up sessions" for working professional librarians and archivists were also suggested to help practicing professionals keep abreast of technical issues.

Research and Development

Alan Lewis (National Archives and Records Administration) suggested that one of our key challenges for many years to come would be to develop reliable methods of copying materials. This is an area that is ripe for research. He noted that equipment to copy the antique formats is becoming increasingly difficult to find and suggested that we may want to explore engaging a company (or several) to actually produce new machines to do this work. Alan also reiterated the value and economic sense of providing temperature- and humidity-controlled storage for both antique and transitional media (magnetic tape) as a way of buying time while we develop cost-effective and faster methods for reformatting.

Several individuals suggested the need for a central research center, (e.g., an equivalent of the Rochester Institute of Technology's Image Permanence Institute) devoted to sound preservation, where research and development could be undertaken. The emphasis here would be on physical materials regarding conservation and playback, and also on computer-related elements that pertain to digital reformatting, technical metadata, and the handling of digital signals and files. These comments raised the question of where in the academic or private sectors an institution of this type might be located and who might fund it. Several participants commented that the computing centers on some campuses might provide the necessary IT setting for digital work. Others mentioned the suitability of existing labs (e.g., labs at Indiana University, Stanford University, LC) as possibilities for developing into focal points for audio preservation research. There was broad consensus regarding the need for diagnostic tools to predict media deterioration and methods for predicting system obsolescence. In this regard, more scientific data on the physical characteristics of media (ageing profiles and deterioration curves) are needed, and more information about commercial products is needed from vendors and manufacturers. Laboratories could address these issues.

Treatment

Sarah Stauderman (Smithsonian Institution) expressed concern about treatment protocols for tapes that are being used by some institutions prior to playback, and commented on the particular lack of research to support, for example, the baking of tapes to consolidate binder chemistry prior to copying. She pointed out that this is only one of a number of practices that require scientific research so that we might better understand their long-term preservation benefits and/or drawbacks. We need to understand the mechanisms of problems such as "sticky shed" (i.e., degradation of the binder material) and "vinegar syndrome" and develop predictive methods for identifying these conditions in our collections. If we know the factors that trigger chemical and mechanical deterioration, we can better understand what we need to do to avoid those conditions. Similar concern was expressed regarding the various untested chemistries that are being used to clean tapes and records. Again, more research is needed to understand the near- and long-term effects of these practices.

Tools

Sam Brylawski (LC) suggested the need for a series of tools that would help sound archivists better do their jobs. One useful tool would be a "Merck manual of maladies" for sound recordings that could perhaps build on the Web site that Sarah Stauderman presented on endangered audio formats. Tied to this might be a series of "good, better, and best" preservation solutions that could be applied to collections of various types and sizes. Other suggestions included the need to develop a preservation matrix of types of sound recordings and their inherent chemical, physical, and system-dependent characteristics, including their projected life expectancies. Interest was also expressed in the creation of a written guide on the care, handling, and storage of audio materials, as well as a guide to current reformatting practices, with a recognition that in the digital domain viable approaches to reformatting are changing on a regular basis and a guide or manual would need to be updatable as approaches change. Participants suggested that, along with guidelines for caring for collections, the guide include a glossary of terms such as "tape squeal," a list of supplies and suppliers, a directory of sound collections, a directory of service providers that do transfer work (this raised the question of how one goes about evaluating lab work and the notion of certifying service providers, which led to an observation that before we had microfilming standards to guide service providers, product quality was uneven), examples of laboratory setups and estimated associated setup and operational costs, and sample statements of work and sample requests for proposals to assist in contracting out for services.

Standards, Guidelines, and Best Practices

It is old news to anyone who has been involved in audio preservation for the past decade or so that there are no hard and fast standards to guide individuals and institutions in the responsible preservation of audio collections. During the analog tape era, many technical standards for calibrating tape machines were developed and common preservation practices (e.g., recording master tapes at 15 inches per second) were developed. As institutions move increasingly into digital recording, preservation guidelines are needed. Participants also suggested that a guide to pertinent standards for reformatting analog and digital would also be a useful tool for collection managers and sound engineers.

Intellectual Control

One of the recurring comments heard throughout the conference and again in the panel session was the need for the field to develop a comprehensive database of what is held across institutions, as well as a registry of items and collections for which digitization is planned, in progress, or completed. This tool would be particularly useful in helping institutions know what exists, what has been preserved, and what is in need of preservation.

Partnerships and Collaborations

Mention was made throughout the panel session of the need for institutions to work collegially and collaboratively on projects of mutual interest and for institutions to do ample research before embarking on pilot projects, because this work may already have been done. For example, training in sound and moving image is underway in several institutions, including New York University, and the University of California, Los Angeles, maintains a program in film preservation. One person suggested that a survey of programs, classes, and laboratories should be carried out to help identify unmet training needs.

Advocacy and Support

Several of the panelists raised the point that there is a need for the audio preservation community to be more visible and to make its voice heard. Tom Clareson (OCLC) suggested that the audio preservation community create an agenda and then develop a strategy for advancing that agenda through advocacy at the local, regional, and national levels. This dialogue led into a discussion of funding strategies, to which Charles Kolb (National Endowment for the Humanities, Division of Preservation and Access) commented that NEH has seen a dramatic rise in applications dealing with sound preservation and that institutions interested in seeking support should consult NEH's newly revised guidelines. On the topic of advocacy, several participants suggested that as a community we should be encouraged to rally historians, scholars, users, and creators of sound materials in support of its preservation.

A suggestion was also made to increase public awareness of what is at risk through a targeted marketing advocacy campaign.

Future Actions

Based on the rich discussion that took place throughout the conference and in the final panel session, there seem to be several immediate actions that stakeholders agree upon, including:

1. Develop undergraduate and graduate level classroom and laboratory curricula for audio preservation. Create fellowship and internship opportunities to build practical skill, knowledge, and ability within institutions.

2. Develop a comprehensive catalog of sound recordings held in public and private institutions and a digital registry of what has been (or will be) preserved.

3. Develop a manual on the care, handling, and storage of audio materials, including sections on analog and digital reformatting, citing pertinent standards and best practices.

4. Develop a series of training workshops for professional librarians and archivists focused on specific topics and involving seasoned engineers.

5. Develop, in collaboration with the stakeholder community, a research agenda and strategy for addressing key technical and scientific issues.

6. Develop an advocacy campaign targeting the public and private sectors focused on what is at risk.

7. Form a coordinating group to advance these actions.

Sound Savings: Bibliography and Resources

Sarah Cunningham, MSIS

Information Specialist, Historical Music Recordings Collection, University of Texas at Austin
Audio Preservation Specialist, Lyndon Baines Johnson Library

Works Cited: Books and Articles

Allen, David Randel. *State of Recorded Sound: Survey of Surveys.* Report prepared by the Communications Office, Inc. for the Council on Library and Information Resources, 2003.

American Folklife Center, Library of Congress. *After the Day of Infamy: "Man-on-the-Street" Interviews Following the Attack on Pearl Harbor.* American Memory: Historical Collections for the National Digital Library. http://memory.loc.gov/ammem/afcphhtml/afcphhome.html.

Association for Recorded Sound Collections, Associated Audio Archives. *Audio Preservation: A Planning Study: Final Performance Report.* Silver Spring, MD: Association for Recorded Sound Collections, 1988.

Association of Recorded Sound Collections. *Rules for Archival Cataloging of Sound Recordings.* 1997.

Babbie, Earl R. *Survey Research Methods.* 2d ed. Belmont, CA: Wadsworth, 1990.

Baker, Holly Cutting. "The Linscott Collection." *Folklife Center News* 3, no. iii (1979): 6–7.

Bishop, Julia C. "The Most Valuable Collection of Child Ballads with Tunes Ever Published: The Unfinished Work of James Madison Carpenter." *In Ballads into Books: The Legacies of Francis James Child*, ed. T. Cheesman, and S. Rieuwerts, 81–94. Bern, Switzerland: Peter Lang, 1999.

Bishop, Julia C., David Atkinson, Elaine Bradtke, Eddie Cass, Thomas A. McKean, and Robert Young Walser, eds. *The James Madison Carpenter Collection Online Catalogue.* 2003. http://www.hrionline.ac.uk/carpenter/index.html.

Borko, H. "Information Science: What is it?" *American Documentation* 19, no.1 (1968): 3–5.

Boyer, Walter E., ed. *Songs Along the Mahantongo: Pennsylvania Dutch Folksongs.* Lancaster, PA: Pennsylvania Dutch Folklore Center, 1951.

Buffington, Albert F., comp. *Pennsylvania German Secular Folksongs.* Breinigsville, PA: Pennsylvania German Society, 1974.

Busha, Charles H. and Stephen P. Harter. *Research Methods in Librarianship: Techniques and Interpretation.* New York: Academic Press, 1980.

Butterworth, W.E. *Hi-fi: From Edison's Phonograph to Quadrophonic Sound.* New York: Four Winds Press, 1977.

Byers, Fred R. *Care and Handling of CDs and DVDs.* Washington, D.C.: Council on Library and Information Resources and National Institute of Standards and Technology, 2003.

Colorado Digitization Program, Digital Audio Working Group. *Digital Audio Best Practices Version 1.2.* May 2003. http://www.cdpheritage.org/resource/audio/std_audio.htm.

Committee on an Information Technology Strategy for the Library of Congress, Computer Science and Telecommunications Board, Commission on Physical Sciences, Mathematics, and Applications, National Research Council. 2001. *LC21: A Digital Strategy for the Library of Congress.* Washington, D.C.: National Academy of Sciences.

Conservation Online. *Preservation of Audio Materials.* http://palimpsest.stanford.edu/bytopic/audio/.

Creswell, John W. *Research Design: Qualitative and Quantitative Approaches.* Thousand Oaks, CA: Sage, 1994.

Dale, Robin, Janet Gertz, Richard Peek, and Mark Roosa. *Audio Preservation: A Selective Annotated Bibliography and Brief Summary of Current Practices.* Chicago: American Library Association, 1998.

"Dialect Collection for Folk Archive." *Folklife Center News* 8, no. 2 (1985): 4–6.

Dickinson, Eleanor. *Revival.* New York: Harper & Row, 1974.

Drewes, Jeanne, and Andrew Robb. "The Use of Handheld Computers in Preservation and Conservation Settings." Paper presented at the annual meeting of the

American Institute for the Conservation of Artistic and Historic Works, Philadelphia, June 2000. Information available at http://www.lib.msu.edu/drewes/Presentation/palmp/survey.html and http://www.lib.msu.edu/drewes/Presentation/palmp/handout0606.doc.

Drott, M. Carl. "Random Sampling: A Tool for Library Research." *College and Research Library News* 30 (1969): 99–125.

Fewkes, Jesse Walter. "A Contribution to Pasamaquoddy Folk-Lore." *Journal of American Folklore* 3 (1890): 257–80.

Gevinson, Alan. "'What the Neighbors Say': The Radio Research Project of the Library of Congress." In *Performing Arts: Broadcasting*, 94–121. Washington, D.C.: Library of Congress, 2002.

Gray, Judith A. and Dorothy Sara Lee, eds. *The Federal Cylinder Project: A Guide to Field Cylinder Collections in Federal Agencies.* Volume 2: *Northeastern Indian Catalog*; *Southeastern Indian Catalog.* Washington, D.C.: American Folklife Center, Library of Congress, 1985.

Griscom, Richard. "Distant Music: Delivering Audio Over the Internet." *Notes* (March 2003).

Kenney, Anne R. and Deirdre C. Stam. *The State of Preservation Programs in American College and Research Libraries: Building a Common Understanding and Action Agenda.* Washington, D.C.: Council on Library and Information Resources, 2002. http://www.clir.org/pubs/reports/pub111/introsum.html.

Kurath, Hans, ed. *Linguistic Atlas of New England.* Providence: Brown University, 1939–43.

ISO Archiving Standards—Reference Model Papers. http://ssdoo.gsfc.nasa.gov/nost/isoas/ref_model.html.

Library of Congress. *Digital Audio-Visual Preservation Prototyping Project.* 23 March 2001. http://lcweb.loc.gov/rr/mopic/avprot/avprhome.html#coll.

Linscott, Eloise Hubbard, ed. *Folk Songs of Old New England.* New York: Macmillan, 1939.

McWilliams, J. *The Preservation and Restoration of Sound Recordings.* Nashville, TN: American Association for State and Local History, 1979.

Maguire, Marsha. "Confirming the Word: Snake-Handling Sects in Southern Appalachia." *The Quarterly Journal of the Library of Congress* 38 (1981):166–79.

Mathews, Max V. *The Technology of Computer Music.* Cambridge, MA: M.I.T. Press, 1969.

Metadata Encoding and Transmission Standard. http://www.loc.gov/standards/mets/.

Metadata Encoding and Transmission Standard: AV Prototype Project Working Documents. http://lcweb.loc.gov/rr/mopic/avprot/metsmenu2.html.

Metadata Object Description Schema. Library of Congress. http://www.loc.gov/standards/mods/.

Michigan State University. *Working Paper on Digitizing Audio for the National Gallery of the Spoken Word and the African Online Digital Library.* http://africandl.org/bestprac/audio/audio.html.

NASA Open Archival Information System (OAIS) Reference Model. http://ssdoo.gsfc.nasa.gov/nost/isoas/.

National Association for the Preservation and Perpetuation of Storytelling. *Best-Loved Stories Told at the National Storytelling Festival.* Jonesborough, TN: National Storytelling Press; Little Rock: August House, 1991.

National Digital Information Infrastructure Preservation Program (NDIIPP) http://www.digitalpreservation.gov/.

National Park Service and National Trust for Historic Preservation. *Save America's Treasures.* n.d. http://www.saveamericastreasures.org/.

National Recording Preservation Board, Library of Congress. http://www.loc.gov/rr/record/nrpb/.

Ottar Johnson, Frédéric Bapst, etc. "Visual Audio: An Optical Technique to Save the Sound of Phonographic Records." *IASA Journal* 21 (July 2003): 38–47.

Pickett, A. G. and M.M. Lemcoe. *Preservation and Storage of Sound Recordings.* Washington, D.C.: Library of Congress, 1959. Reprint, Association of Recorded Sound Collections, 1994.

Powell, Ronald R. *Basic Research Methods for Librarians.* 2d ed. Norwood, NJ: Ablex, 1991.

Preservation Metadata and OAIS Information Model. Research Libraries Group and OCLC (Online Computer Library Center). http://www.oclc.research/pmwg/

PRESTO (Preservation Technology for European Broadcast Archives) project organized by broadcasters in Europe. http://presto.joanneum.ac.at/index.asp

Read, O. and W. L. Welch. *From Tin Foil to Stereo: Evolution of the Phonograph.* Indianapolis, IN: Howard W. Sams & Co. 1959.

"RIT Studies Increasing Shelf Life for History Stored on Tape." Rochester Institute of Technology press release, June 25, 2003.

Seubert, David. "Designing and Managing an Audio Preservation Program." Paper presented at the annual meeting of the Association for Recorded Sound Collections, Philadelphia, May 2003.

Smith, Abby. "Background for 14 March meeting." E-mail to Connie Brooks. 1 March 2003.

———. "CLIR to Survey Audio Collections in Academic Libraries." *Council on Library and Information Resources* 32 (March/April 2003).

———. "Finding Our Voice: Survey Highlights Barriers to Access of Audio Collections." *Council on Library and Information Resources* 40 (July/August 2004). http://www.clir.org/pubs/issues/issues40.html#voice

Smith, Jimmy Neil. "Storytelling Collection Comes to the Library of Congress." *Folklife Center News* 23, no. 3 (2001): 3–5.

Spring, M. *Electronic Printing and Publishing: The Document Processing Revolution.* New York: Marcel Dekker Inc., 1991.

Tedlock, Dennis, trans. *Finding the Center: The Art of the Zuni Storyteller.* 2nd ed. Lincoln: University of Nebraska Press, 1999.

Thompson, James. "The End of Libraries." *The Electronic Library* 1, no. 4 (October 1983): 245.

University of California, Regents. "How Much Information." http://www.sims.berkeley.edu/research/projects/how-much-info/summary.html.

Van Bogart, J. *Magnetic Tape Storage and Handling: A Guide for Libraries and Archives.* Washington, D.C.: National Media Laboratory and Council on Library and Information Resources, 1995.

Van Praag, P. *Evolution of the Audio Recorder.* Waukesha, WI: EC Designs, Inc., 1997.

Ward, A. *A Manual of Sound Archive Administration.* Hants, England: Gower Publishing, 1990.

Warren Jr., Richard. "Handling of Sound Recordings." *ARSC Journal* 25 (Fall 1994): 139–62.

Yoder, Don. *Pennsylvania Spirituals*. Lancaster, PA: Pennsylvania Folklife Society, 1961.

Works Cited: Web Sites

AES Audio Engineering Society http://www.aes.org/

ARSC Association of Recorded Sound Collections http://www.arsc-audio.org/

Art Shifrin http://www.Shifrin.net/

Audio Preservation Bibliography and Web Reference http://palimpsest.stanford .edu/bytopic/audio/

Consumer Audio http://mhintze.tripod.com/audio/default.htm

Cutting Corporation http://www.cuttingarchives.com/head/faq.html

DVD FAQ http://www.dvddemystified.com/dvdfaq.html

Edison Cylinders http://www.nps.gov/edis/pr_loc_rec_020103.htm

Edison Museum http://www.edisonnj.org/menlopark/birthplace/northamerican phonograph.asp

History of Recorded Sound Technology http://www.recording-history.org/HTML/ start.htm

Internet Museum of Flexi / Cardboard / Oddity http://www.wfmu.org/MACrec/ index.html

Library of Congress Preservation of Sound Recordings FAQ http://lcweb.loc.gov/ preserv/care/record.html

Optical Storage Technology Association http://www.osta.org/

Recording Technology History http://history.acusd.edu/gen/recording/notes.html

Richard Hess http://www.richardhess.com

Sound Reproduction R & D Home Page. http://www-cdf.lbl.gov/~av/

Steve Smolian http://www.soundsaver.com

VidiPax http://www.VidiPax.com

Other Resources

Association for Recorded Sound Collections (ARSC). *Education and Training in Audiovisual Archiving and Preservation.* 2004 http://www.arsc-audio.org/ETresources.html

Audio Engineering Society Historical Committee. AES http://www.aes.org/aeshc/

The British Library. "The British Library Sound Archive." 2004 http://www.bl.uk/collections/sound-archive/artefacts.html

Brylawski, Samuel. "Preservation of Digitally Recorded Sound."*Building a National Strategy for Preservation:Issues in Digital Media Archiving.* Washington D.C.: Council on Library and information Resources, 2002.

Child, Margaret. *Directory of Information Sources on Scientific Research Related to the Preservation of Sound Recordings, Still and Moving Images, and Magnetic Tape.* Commission on Preservation and Access. 1993. http://www.clir.org/pubs/reports/child/child.html

"Collections Care and Conservation." *Preservation Department of the Library of Congress.* Library of Congress. http://lcweb.loc.gov/preserv/pubscare.html#mag

Council on Library and Information Resources (CLIR). *Folk Heritage Collections in Crisis.* http://www.clir.org/pubs/reports/pub96/contents.html

Dale, Robin, Janet Gertz, Richard Peek, and Mark Roosa. *Audio Preservation: A Selective Annotated Bibliography and Brief Summary of Current Practices.* Chicago, IL: American Library Association, 1998. http://www.ala.org/ala/alctscontent/alctspubsbucket/webpublications/alctspreservation/audiopreservatio/audiopres.pdf

Farrington, Jim. "Preventive Maintenance for Audio Discs and Tapes." *Notes: Quarterly Journal of the Music Library Association* 48, no. 2 (1991): 437–45.

Fells, Nick, Pauline Donachy, and Catherine Owen. *Creating Digital Audio Resources: A Guide to Good Practice.* Oxford: Oxbow, 2002.

Gibbs, John R. "Audio Preservation and Restoration." University of Washington Music Library. http://www.lib.washington.edu/music/preservation.html

Grammy.com. *Preservation.* Grammy Gateway. 2004. http://www.grammy.com/gateway_preservation.aspx#Sound

International Association of Sound Archives. "The Safeguarding of the Audio Heritage: Ethics, Principles and Preservation Strategy." International Association of Sound and Audiovisual Archives: Version 2, September 2001. http://www.iasa-web.org/iasa0013.htm

National Archives and Records Administration. *Preservation Reformatting: Digital Technology vs. Analog Technology.* Preservation, 2003. http://www.archives.gov/preservation/conferences/preservation_conference2003.html

Nelson-Strauss, Brenda. "Preservation Policies and Priorities for Recorded Sound Collections." *Notes: Quarterly Journal of the Music Library Association* 48, no. 2. (1991).

"Recording Technology History". *Recordist.* http://www.recordist.com/

Schoenherr, Steve. *"Recording Technology History."* notes revised Feb. 16, 2004. http://history.acusd.edu/gen/recording/notes.html

Schüller, Dietrich. "Ethics of Preservation, Restoration, and Reissue of Historical Sound Recordings." *Journal of the Audio Engineering Society* 31, no. 12: 1014–16

St. Laurent, Gilles. *The Care and Handling of Recorded Sound Materials.* Washington, D.C.: Commission on Preservation and Access, 1996.

Syracuse University. *Belfer Audio Laboratory & Archive Audio Archiving: Current Issues and Selected Readings.* http://libwww.syr.edu/information/belfer/bibliogr.htm

Van Bogart, John W.C. *Magnetic Tape Storage and Handling: A Guide for Libraries and Archives.* Minneapolis, MN: Commission on Preservation and Access and the National Media Lab, June 1995.

United States Copyright Office. *Copyright.* 2004. http://lcweb.loc.gov/copyright